BASIL COPPER

A LIFE IN BOOKS

BASIL COPPER

A LIFE IN BOOKS

Compiled and Edited by

STEPHEN JONES

PS Publishing 2008

First published in February 2008 by PS Publishing Ltd. by arrangement with the author.

A Smith & Jones Book

FIRST EDITION

ISBN 978-1-905834-98-3 (Jacketed hardcover)
ISBN 978-1-905834-97-6 (Trade hardcover)

A CIP catalogue record for this is available from the British Library.

Printed and bound in Great Britain by Biddles Ltd.

PS Publishing Ltd.
Grosvenor House
1 New Road
Hornsea HU18 1PG
England

e-mail: *editor@pspublishing.co.uk*
Internet: *http://www.pspublishing.co.uk*

Many of Basil Copper's original papers are preserved and available for study through the 'Contemporary Collections' repository of the Howard Gotlieb Archival Research Center at Boston University, USA.

CONTENTS

INTRODUCTION: RECOLLECTIONS OF BASIL COPPER by Stephen Jones 11

FOREWORD by Basil Copper 14

BASIL COPPER: CRIMEWRITER & MASTER OF THE MACABRE by Richard Dalby 17

BASIL COPPER: A WORKING BIBLIOGRAPHY by Stephen Jones 30

A. Macabre & Supernatural Novels 31

REMEMBERING AUGUST DERLETH & ARKHAM HOUSE by Basil Copper 36

B. Macabre & Supernatural Collections 39

THE GAME'S AFOOT! AUGUST DERLETH & SOLAR PONS by Basil Copper 43

C. The "Solar Pons" Series 49

AUGUST DERLETH: A GIANT REMEMBERED by Basil Copper 56

D. The "Mike Faraday" Series 67

RAP-SHEET by Basil Copper 82

DIAMONDS ARE FOR NEVER by Basil Copper 85

E. Other Books 104

IN THE FOOTSTEPS OF DRACULA by Basil Copper 107

F. Non-Fiction 111

THE CURSE by Basil Copper 117

PROFESSOR MAHMOUD by Basil Copper 119

G. Short Stories & Novellas 131

PETER HAINING: A TRIBUTE by Basil Copper 158

THE BLACK COFFIN by Basil Copper 159

H. Media Adaptations 163

COUNT MAGNUS 167

I. Unpublished Works 263

DARKNESS, MIST & SHADOW—THE LANDSCAPE OF FEAR by Basil Copper 267

ACKNOWLEDGEMENTS

Special thanks, first and foremost, to Basil and Annie Copper, Peter Crowther, Michael Marshall Smith, Richard Dalby, Val and Les Edwards, Randy Broecker, Mike Ashley, Mandy Slater, Jo Fletcher, Ian Drury, Peter Haining, R. Dixon Smith, David Sutton, Ramsey Campbell, Nancy A. Collins, Michel Parry and Amanda Foubister, for all their help and advice in preparing this bibliography.

–SJ.

I am deeply grateful to my close friend, Stephen Jones, for the care and scholarship he has brought to the preparation of my Bibliography, in the course of which he has dredged up material which I had completely forgotten.

–BC.

ILLUSTRATIONS

DEDICATION

This one is in memory of my beloved parents, Ethel Maud Copper and Edgar Gerald Copper, both of whom sadly died long before they could ever see any of my eighty-two books in print.

—BC.

INTRODUCTION: RECOLLECTIONS OF BASIL COPPER

I FIRST MET BASIL COPPER more than twenty years ago. Neither of us really remembers much about the encounter, which is probably a blessing.

I had almost certainly already read his story 'The Spider' in Herbert van Thal's *The Fifth Pan Book of Horror Stories* (1964) while still a teenager in my school's library, and I had admired Rod Serling's adaptation of his story 'Camera Obscura' on TV's *Night Gallery* in the early 1970s; but at the time we first met I knew him best for his classic tale 'Amber Print', which I had discovered in Peter Haining's anthology *Dr. Caligari's Black Book* (1968).

It was 1977, and I was attending Fantasycon 3 in Birmingham, England. Basil was the Guest of Honour that year and, according to the programme booklet that I have before me (for which, incidentally, I contributed the cover illustration), he gave an hour-long lecture on the Saturday afternoon. I know I was there (I used to attend all the programme items at conventions in those days—I was *keen*), but nearly three decades later the only memory of the entire weekend that I can recall is being in the same room party as Basil on the Saturday night, surrounded by crates of beer and a well-known book dealer dancing around with a pair of ladies knickers on his head!

I believe that it was more than a decade before Basil was sufficiently recovered from the ordeal to attend another fantasy convention.

In the meantime I had become a big fan of his work. My shelf of Basil's books now included his two excellent collections for Arkham House, *From Evil's*

by Stephen Jones

Pillow (1973) and *And Afterward, the Dark* (1977), and the novels *Necropolis* (1980) and *The House of the Wolf* (1983). I had also managed to pick up a copy of his earlier novel, *The Curse of the Fleers* (1976), and our mutual friend at Arkham House, James Turner, had sent me some of Basil's Solar Pons collections as they were published in paperback by Pinnacle Books during the late 1970s.

I was going to say that our next meeting was under more sober circumstances, but that is not entirely true. It was in 1988, at the lunchtime opening of a new crime and mystery bookstore in London's West End named after one of his Mike Faraday novels. In the course of admiring the new shop fittings and drinking the indifferent white wine usually served at these occasions, we got to talking, and I think Basil was genuinely surprised to meet someone who was such a fan of his work and who could quote title and publication date from memory (at least before we began drinking). Whatever the reason, when the party started to wind down, I suggested we move on to a small all-day drinking club I happened to know in Soho. Basil readily agreed. Over several hours—and many, *many* more glasses of wine—we talked about the art of writing and bemoaned the state of the horror field (some things never change), and when we finally reeled out into the dark night to make our bleary journeys back to our homes, a lasting friendship had been forged over alcohol and good conversation.

In the decade since then we have met up on numerous occasions—often at conventions and gatherings on both sides of the Atlantic, sometimes sharing the same panel discussions and signing sessions, and I had the honour of interviewing him in depth about his fascinating life and career at Fantasycon XX in 1996. I have been his guest, and that of his beautiful French wife Annie, many times at their charming home in Kent. As an editor, I have been proud to work with Basil on numerous anthologies and non-fiction books, and (with his help) I have even managed to fill almost all of those gaps in my Basil Copper shelf as the years have passed.

As a result, I finally decided to put all that material to good use. With a great deal of help from Basil himself—who went through his own files and personal library to supply previously unpublished or rare items—I have attempted to compile a bio-bibliography that would be beneficial to dedicated collector's of Basil's work, along with those readers who are only now discovering his extensive and eclectic output.

My thanks to Richard Dalby, who very kindly updated his excellent article about Basil from *Book and Magazine Collector* for this volume, Michael Marshall Smith for his usual immaculate design work, and Peter Crowther of PS Publishing, for agreeing to publish this book almost as soon as I mentioned it to him. And last, but not least, to Basil and Annie, for opening up their lives to me, and inviting me into their home on many occasions and making me feel so welcome. Without their tireless support and enthusiasm for this project, it would have been all the poorer.

For me, creating this book has been an opportunity to produce a useful *aide-mémoire* as well as promoting Basil's many books to a wider audience. I hope it does both the author and his wide body of works the justice they deserve.

With the passing of years, my memory of our first meeting may now be getting a little hazy, but to any new readers who are just beginning to assemble a collection of his books, I can guarantee that like me, you will find Basil Copper's fiction distinctly . . . *unforgettable*!

Basil Copper and Stephen Jones, circa 1990s

FOREWORD

IN INTRODUCING THIS VOLUME I hope I may be forgiven for making a few observations on the genre.

The great masters of the past like Poe, Bierce, Hawthorne, de Lisle d'Adam and those of more recent times, such as Blackwood, M.R. James, Lovecraft and others too numerous to mention, rely heavily on the Gothic for their effects.

This is inevitable in stories where atmosphere must count for so much, but many of the most effective and terrible moments in the finest tales of the macabre occur in settings of bright sunshine, where all promises to be well.

Possibly the most telling of all such moments occurs in Henry James' 'Turn of the Screw', with the appearance of Peter Quint on the riverbank. This has the right, dry-mouthed terror almost impossible to convey without re-reading the whole of that masterpiece.

The Irish master, Le Fanu, too, was almost unapproachable in these effects and I should rate 'Uncle Silas' a very high place among the findings of that fatuous and impossible occupation, the compilation of a "Ten Best" of this, that and the other.

In my writing, I have tried to combine both worlds and the reader will find tales of light and shadow which will, I hope, cause him or her to cast an uneasy glance over the shoulder from time to time.

But the medium, perhaps more than any other, does demand the highest degree of co-operation from the reader. He must, in fact, believe before he even

by Basil Copper

opens the volume; given that and a reasonable degree of skill on the part of the author, he will be transported into a malign and enchanted world in which all things are possible.

Despite the events of the later decades of the twentieth century, an age of poor ideals, shabby politics and dubious plastics, the market for tales of mystery and horror continues to increase and so will authors provide new stories to fill that insatiable demand.

Perhaps the modern reader reads to escape from an epoch of even more plausible horror; certainly, in the atomic age, he has even greater terrors to live with than had mediaeval man with his dragons and powers of darkness.

Whatever the reason, he must be, as I said before, a believer in the impossible made possible. Otherwise, he will not extract full enjoyment from books such as mine.

But for the reader who relishes the macabre, there will be, I believe, a genuine shudder in the collections of stories I have gathered together.

At least I hope so. It will not have been for want of trying.

—Basil Copper
Rouvray, Cote d'Or

Dark Detectives *launch, London, February 2000.*
Back row (LtoR): Basil Copper, Jay Russell, Brian Lumley, Stephen Jones.
Front Row (LtoR): Kim Newman, Les Edwards, Randy Broecker.

PRAISE FOR BASIL COPPER

"An outstanding British writer in the genre."

—*August Derleth*

"He beguiles the mind as he lures the imagination beyond the outposts of reality."

—*Donald Wandrei*

"One of the last of the great traditionalists of English fiction."

—*Colin Wilson*

"Britain's leading purveyor of the macabre."

—*Peter Haining*

"He can handle the macabre with the best . . . in the same class as M.R. James and Algernon Blackwood."

—*Michael and Molly Hardwick*

"One of the twentieth century's masters of horror and the macabre."

—*R. Dixon Smith*

"Copper is undoubtedly one of the best modern writers on the basic theme that might well be called the 'Dracula menace'."

—*Richard L. Tierney*

"A new master of terror"

—*Professor Edward Wagenknecht*

"Basil Copper is, I think, the best writer in the genre since H.P. Lovecraft."

—*Frederick Shroyer, Literary Editor,*
Los Angeles Herald-Examiner

BASIL COPPER: CRIMEWRITER & MASTER OF THE MACABRE

ACCLAIMED AS ONE OF BRITAIN'S greatest masters of macabre and Gothic fic-tion, Basil Copper has also achieved equally popular success with no less than fifty-two "hard-boiled" detective thrillers, all featuring Mike Faraday, as well as around thirty "Sherlock Holmes" pastiches about the great Sherlockian disciple, Solar Pons.

Colin Wilson perhaps came nearest to defining his genius when he described him as "one of the last great traditionalists of English fiction", following in a long and distinguished line of writers that includes Conan Doyle, Wilkie Collins and Sheridan Le Fanu.

Copper has received many honours in recent years. In 1979, the Mark Twain Society of America elected him a Knight of Mark Twain for his "outstanding contribution to modern fiction", while the Praed Street Irregulars have twice honoured him for his work on the "Solar Pons" series. He has been a member of the Crime Writers' Association for over thirty years, serving as chairman in 1981-82 and on its committee for a total of seven years.

Basil Copper was born in London on 5th February 1924, but moved with his family to Kent, where he attended a grammar school. However, the school failed to satisfy his wide interests, which included classical music and collecting records. He haunted bookshops, libraries and concert halls, and read voraciously—anything from the classics to the macabre works of M.R. James and

by Richard Dalby

Edgar Allan Poe, both of whom were to have great influence on his later work.

While training as an apprentice journalist, he attended a local commercial college, where, "in addition to bookkeeping, economics and other subjects, I found the only things of any use were shorthand and touch-typing, which have remained with me all my life".

Journalism was a demanding profession in the first year of the war, when many established reporters had been conscripted. At the age of seventeen, Copper was put in charge of a county newspaper branch office, while also serving in the Home Guard.

He then joined the Royal Navy, in which he served for four years. After completing a course at the Glasgow Wireless College, he served in Light Coastal Forces—mainly in torpedo boats and gunboats—at Newhaven, Portsmouth and Portland, surviving many encounters with the enemy, in the course of which he was bombed, mined and machine-gunned from the air. "But I bore a charmed life and on one occasion I had a bullet cut through the trousers of my uniform, without cutting the skin, while others were killed around me."

Copper also took part in the D-Day landings: "I was a radio operator on board a motor gunboat flotilla off the Normandy beaches at D-minus one, which escorted units in the first wave of landing craft ashore. During the operations we lost three of our flotilla out of six, mostly to acoustic mines, and were at sea for six weeks and unable to wash! But my luck held and, after going on survivor's leave to my home in the Tunbridge Wells area, I found that my new flotilla had gone on a disastrous operation off the coast of Holland, and that the entire flotilla had been lost, the crews either killed or captured by the Germans. An old friend, who met me at the gates of Portsmouth Barracks at the time, almost fainted as I had apparently been reported killed!"

He subsequently spent a total of two years on radio stations in Egypt, Malta and Gibraltar. Following demobilisation, he returned to journalism, eventually becoming editor of a Kent county newspaper, during which time he contributed to three national newspapers: the *London Evening Standard*, the *Evening News* and *The Star*. From the beginning of the war, he had also contributed to the London Times. In 1960, he married Annie Renée Guerin, a French girl whom he had met while she was learning English in Kent.

Copper had always been fascinated by the Gothic and the supernatural, and his earliest stories were in these genres. As it happened, his first two published tales were destined to become his best known. These were 'The Spider', a story about a man who is haunted by his worst phobia, and 'Camera Obscura', which appeared in the *Fifth* (1964) and *Sixth* (1965) *Pan Books of Horror Stories* respec-

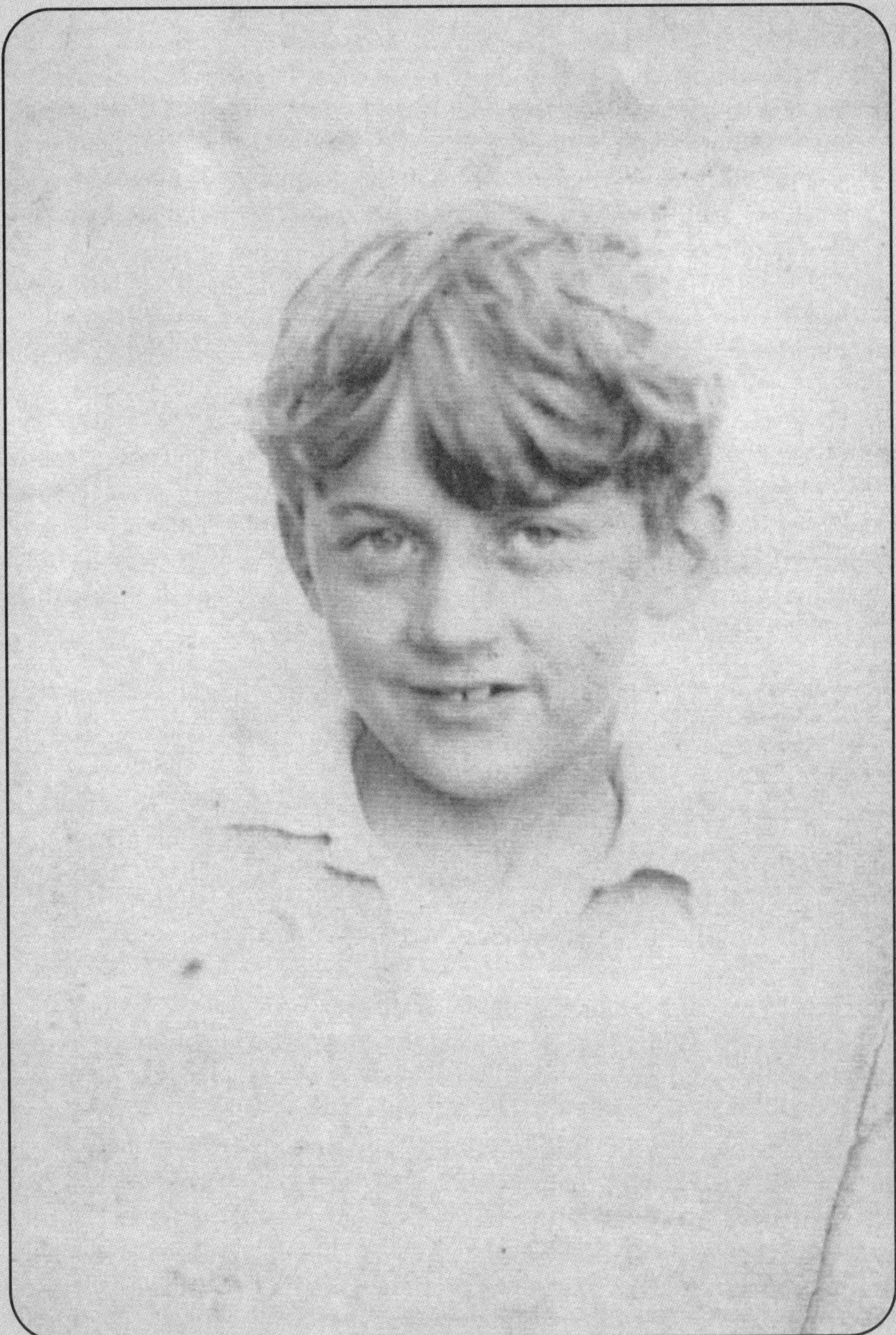

The future author, aged around twelve years old

tively. Six years later, the latter was dramatised for Rod Serling's TV series, *Night Gallery*, with Ross Martin in the lead role.

Copper recalls how his first novel took shape while he was working in the newspaper office: "One day my editorial staff were talking about modern crime novels and one or two said they thought I could do better. It started as a joke, but I set out to write a tongue-in-cheek thriller in the Dashiell Hammett/ Raymond Chandler mode called *The Dark Mirror*. The reporting staff were impressed as the chapters rolled, and when the novel was completed I sent it to thirty-two publishers without any result, until one only replied to the effect that the book was very professionally written but three times too long!

"I had four goes at cutting it down, and it was published by Robert Hale in 1966. I little knew then that the saga would run to fifty-two volumes, and that the books would be translated in paperback into German, Danish, Swedish, Italian, French, Spanish, Norwegian and Czech, to name but a few. Today they are newly appearing in large print editions for libraries and have been recorded for in-flight aircraft entertainment and into talking books for the blind."

Copper's first book, *The Dark Mirror*, launched his very popular and long-running series of hard-boiled thrillers starring Mike Faraday. These appeared at the rate of two or more books a year until the series ended in 1988.

Los Angeles investigator Mike Faraday is a deliberate and very successful homage to Raymond Chandler's Philip Marlowe. Bearing in mind that Copper had never been to California, his descriptions of the state's sleazy hotels, bistros and burger bars are wonderfully convincing. Regular characters—from Faraday's secretary, Stella, to his old powder-blue Buick—give the series a strong sense of continuity.

As both narrator and protagonist, Faraday comes over as a wryly humorous, honest and very positive hero, with many endearing idiosyncrasies. He is particularly fond of quoting from both Robert Herrick and Palgrave's *Golden Treasury*.

Several American authorities have paid glowing tributes to Copper's "Mike Faraday" novels, pointing out that the books' atmosphere and authenticity cannot be faulted, and comparing him to Raymond Chandler. Eminent Sherlockians, Michael and Mollie Hardwick, were also champions of Copper's work over many years. If only the "Mike Faraday" stories had been better-known in the United States, they could have been the basis for a marvellous television series along the lines of *Cannon* and *Columbo*.

As soon as it became clear that the books were to become a series, the jackets were individually numbered and were given their own logo: a Smith &

Wesson .38 special. This replaced the earlier "smoking gun" motif used by Hale.

A great bonus with all the "Mike Faraday" first editions are the dust-jackets, which boast brilliant photomontages designed by the wonderfully imaginative Colin Andrews. One of his best is for *Scratch on the Dark* (1967), which shows three bullets and a photo-album displaying colour cigarette cards of Shirley Temple, Clark Gable and Jean Harlow, with a blank space for the murdered star labelled "Zara Fayne". The whole picture is covered with a looped strip of 35mm film.

The Hook (1984) shows the intimidating image of Santa Claus with gaping skulls as eyes, and all of the other jacket designs in the "Faraday" series have similarly stunning montages.

Basil Copper (left) aboard the Royal Navy's Gunboat 317

I believe that Colin Andrews is one of the most underrated jacket artists working today. An illustrated volume containing all his photo montage designs, including the fifty-two "Mike Faraday" novels and his six other Copper jackets, would admirably display his great talent.

Many of the fifty-two "Faraday" novels have been published in Thorpe large-print editions, as well as paperback. Chivers Press have reissued several of the books–including *The Dark Mirror*–in their hardback C.W.A. (Crime Writers' Association) "Black Dagger" series. Copper has also supplied forewords for some of the other books in this series, including Leslie Charteris' *The Saint Goes West.*

Copper's first collection of supernatural horror tales, *Not After Nightfall,* was published in paperback by New English Library in June 1967. As well as his first two stories, 'Spider' and 'Camera Obscura', this volume also contained 'The Cave', 'The Grey House', 'Old Mrs. Cartwright', 'Charon', 'The Great Vore' and 'The Janissaries of Emilion', most of which have become regular anthology favourites. In his foreword to this book, the author rightly stated that "despite the events of these later decades of the twentieth century, an age of poor ideals, shabby politics and dubious plastics, the market for tales of mystery and horror continues to increase and so will authors provide new stories to fill that insatiable demand".

Copper published three further horror collections over the next thirteen years: *When Footsteps Echo* (1975), *Here Be Daemons* (1978) and *Voices of Doom* (1980). These were all issued by Robert Hale in the UK, and simultaneously in the United States by their sister company, the New York-based St. Martin's Press. All have wonderfully spooky jacket designs by the redoubtable Colin Andrews. These three hardbacks are now very elusive, although the paperback edition of *Here Be Daemons* (Sphere, 1981) is still easy to find.

Most of the stories from the out-of-print *Not After Nightfall* reappeared in the Hale collections, alongside others written in the late 1960s and 1970s for various anthologies. These included 'The Recompensing of Albano Pizar' (1966), which was dramatised in Radio Four's *Man in Black* series; and Copper's Gothic classic of mounting insanity, 'The Knocker at the Portico' (1971), which I was delighted to include in my anthology, *Twelve Gothic Tales* (OUP, 1998). His Halloween horror story, 'The Candle in the Skull', was read on BBC Radio Four on 31st October 1997.

In 1974, Copper wrote a feature-length film script based on M.R. James's story, 'Count Magnus'. This was described as "a terrific read" by Lawrence Gordon Clark, director of BBC TV's annual 'Ghost Story for Christmas', but the BBC was on a strict economy drive in that year, and instead broadcast a cheaper adaptation of Dickens' 'The Signalman', starring Denholm Elliott. Copper's excellent script for 'Count Magnus' is still awaiting production.

Always a keen admirer of H.P. Lovecraft and other classic writers of the genre, Copper regularly corresponded with the Arkham House publisher (and Solar Pons creator), August Derleth, who in turn was extremely enthusiastic about Copper's own horror fiction. Derleth was determined to introduce him to American readers, but it was only after his sudden death that Arkham House published two collections of Copper's stories: *From Evil's Pillow* (1973), with a fine dust-jacket illustration of the three statuesque 'Gossips' by Frank Utpatel; followed by *And Afterward, the Dark* (1977), with a dust-jacket picture of 'The Janissaries of Emilion' by Stephen E. Fabian. This latter volume was introduced by the eminent literary critic Edward Wagenknecht.

From Evil's Pillow was the runner-up in the "Short Story" category at the first World Fantasy Convention. It contained not only 'The Gossips'—a story of demonic power unleashed in modern London—but also 'Amber Print', which offered a dire warning to all collectors of silent films, especially *Das Kabinett des Doktor Caligari*! Copper has been for many years an avid collector and historian

In 1960, Basil Copper married Annie Renée Angele Geurin, a french student he had met while she was studying English. And, as they say, they lived happily after!

of silent film classics and early talkies, and is the founder of The Tunbridge Wells Vintage Film Society.

While making his debut with Arkham House, Copper received another American commission to write two "Phantom" novels on behalf of Lee Falk. *The Slave Market of Mucar* and *The Scorpion Menace* were both published in 1972 by Avon Books, alongside thirteen other titles in this paperback series. (Copper did *not* write the first volume in this series, *The Story of the Phantom* [1972], as stated in Allen Hubin's *Crime Fiction* and other reference works.)

The author at work in his flat, circa 1960s.

The first non-"Faraday" novel by Copper to be published by Robert Hale was *The Great White Space* (1974). Dedicated to H.P. Lovecraft and August Derleth—"Openers of the Way"—this thrilling narrative (with a suitably unsettling jacket design by Colin Andrews) is not only inspired by Lovecraft's "Cthulhu Mythos", but is also firmly rooted in the tradition of Conan Doyle's greatest fantasies, particularly *The Lost World*, in which a journalist, Malone, accompanies Professor Challenger to a remote plateau.

Like Doyle's Malone, the narrator, Plowright, joins Professor Clark Ashton Scarsdale's ill-fated Great Northern Expedition, which ventures into realms of horror in the centre of the Earth. They penetrate the Plain of Darkness and a sinister gallery, in which thousands of giant grasshopper-like creatures are preserved in jars, before encountering the ultimate horror in the Great White Space itself. And what better place to stand in for the "Great White Space" on the book's dust-jacket than London's South Bank Centre, seen through a distorting fish-eye lens?

This novel, issued in the States a year later by St. Martin's Press, prompted the respected critic, Dr. Frederick Shroyer, to describe Copper as "the best writer in the genre since H.P. Lovecraft". Shroyer later wrote: "Copper's versatility is such that he is not confined to the Lovecraft genre. He can move into Poe country like a seasoned guide, and he can evoke the horror that crouches in the human mind when by revenge possessed, as easily as he can mine into the supernatural crypts for the substance of his stories."

At around the same time, Hale commissioned Copper to write his only two "non-fiction" books: *The Vampire: In Legend, Fact and Art* (1973) and its companion, *The Werewolf: In Legend, Fact and Art* (1977). The first title was especially popular, being reissued in paperback and translated into German, but *The Werewolf* has never been reprinted. Also in the vampiric genre, Copper wrote the sleeve notes for the 1974 LP of *Dracula*, narrated by Christopher Lee.

After the success of *The Great White Space,* Copper went on to write three excellent Gothic novels in quick succession. *The Curse of the Fleers* (Harwood-Smart, 1976; St. Martin's Press, 1977), is set in Victorian Dorset, and describes an ancient feud between the Fleer and the Darnley families. The colourful dust-jacket painted by Dave Austin is embellished by a huge grinning skull, with a bloody teardrop. The original text was heavily cut and revised, and the novel has never been published in its definitive form.

Necropolis (1980), illustrated by Stephen E. Fabian, was the most successful book in the first forty-one years of Arkham House's history, with all 4,050 copies selling very quickly. To satisfy demand, the firm printed a further 1,539 copies in 1981—making *Necropolis* the first non-Lovecraft Arkham House title to achieve a second impression.

This macabre mystery, set in Victorian England, was actually the first straight Gothic novel to be published by Arkham House, and marked an important deviation from their familiar output of weird and supernatural horror.

In *The House of the Wolf* (Arkham House, US, 1983), American professor John Coleridge journeys to the Hungarian village of Lugos—birthplace of Bela Lugosi—and the nearby Castle Homolky. Like its predecessor, this book featured a fine dust-jacket—as well as forty-three atmospheric chapter-heading illustrations—by Stephen E. Fabian. This novel was not published in Britain until 2003 (by Sarob Press), although *Necropolis* was issued in the UK by Sphere in 1981.

Sphere also published the only edition of Copper's *Into the Silence* in 1983. Following in the tradition of *The Great White Space* and Conan Doyle, this exciting novel is set in the 1920s, and features another Challenger-like scientist, Professor Deems, who directs his team to burrow through solid granite into a cavern lying under a Cornish village "a journey beyond darkness into total, screaming terror!", to quote the cover blurb.

August Derleth had begun writing his long series of stories about Solar Pons (derived from the Latin for "bridge of light") in the late 1920s after receiving a letter from Sir Arthur Conan Doyle declaring that there would be no more

"Sherlock Holmes" stories. Pons was closely modelled on Doyle's Great Detective–he lived at 7B Praed Street, near Paddington Station, with his Watson counterpart, Dr. Lyndon Parker, and his ever-faithful landlady, Mrs Johnson–and he eventually became the most enduring of *all* Sherlock Holmes' Successors.

Derleth created a new imprint–Mycroft & Moran–especially for the "Solar Pons" series, which ran to eight volumes, from *"In Re: Sherlock Holmes" – The Adventures of Solar Pons* (1945) to the posthumously-published *Chronicles of Solar Pons* (1973). These appeared at a time when Sherlockian pastiches were still very thin on the ground, and proved to be very popular.

Unfortunately, Derleth never visited Britain, and did very little research for the stories, with the result that they are rather less authentic than Copper's "Mike Faraday" books. In their mammoth *Catalogue of Crime*, Jacques Barzun and Wendell Hertig Taylor are scathing about Derleth's mistakes: "In several instances," they write, "the author's lack of familiarity with English ways and terminology is painful."

When Copper pointed out a number of these errors to James Turner, Derleth's successor at Arkham House, Turner commissioned him to revise and edit the entire Pontine canon. This mammoth task, a real labour of love, took Copper nearly eighteen months and involved correcting around 2,000 errors in Derleth's seventy short stories and one novel.

These revised texts were published by Arkham House in 1982 as *The Solar Pons Omnibus*, a two-volume, slip-cased edition, beautifully illustrated by Frank Utpatel and with a foreword by Robert Bloch. Published at $39.95, this classic set is now in great demand, fetching upwards of £100.00.

In the late 1970s, Turner asked Copper if he would like to continue the "Pons" series himself. The author eagerly accepted, producing six collections, which were enthusiastically received by Turner. All of the stories were novella-length (sixty to seventy-five pages), which gave them more depth and scope than Derleth's much shorter tales, and they were full of memorably macabre touches.

Unfortunately, Arkham House never published these stories, although the New York-based firm, Pinnacle, did issue four paperback collections–*The Dossier of Solar Pons, The Further Adventures of Solar Pons, The Secret Files of Solar Pons* (all 1979) and *Some Uncollected Cases of Solar Pons* (1980). These were all very badly "revised" by in-house editors, who, as Copper recalls, "completely destroyed the 1920s British atmosphere by rewriting and inserting crass and stupid alterations".

Academy Chicago eventually published only two of these books, *The Dossier*

The youthful bookworm in a corner of his bedroom during the early days of World War II

of Solar Pons and *The Further Adventures of Solar Pons* (both 1987), returning to the author's original "definitive" texts, but the other two are still awaiting revival in similar definitive state.

After nearly twenty years in limbo, two previously unpublished collections (like the others, all written in the late 1970s) were issued by the Minneapolis-based firm of Fedogan & Bremer: *The Exploits of Solar Pons* (1993) and *The Recollections of Solar Pons* (1995).

Superbly illustrated by Stefanie Kate Hawks, both volumes were produced in ordinary and deluxe editions, the latter limited to 100 slip-cased copies, signed by the author and artist. The limited editions went out of print very quickly and are now collector's items, selling for up to £150.00 in fine condition.

Fedogan & Bremer also published Copper's latest Gothic thriller, *The Black Death* (1991), which is set in the fog-shrouded wilds of Dartmoor. Again illustrated by Stefanie Kate Hawks, this book was similarly issued in both trade and deluxe editions—although the former would also be considered "deluxe" by most collectors, as the production values of this company are extremely and painstaking.

Basil Copper with artist Stefanie Kate Hawks, 1994 World Horror Convention, Phoenix. Arizona

In March 1999, Fedogan & Bremer issued Copper's long-awaited new collection of macabre stories, *Whispers in the Night*, richly embellished with illustrations by Stephen E. Fabian. Among the stories in this volume were the memorable novella, 'Wish You Were Here' (commissioned for my anthology *Horror for Christmas* in 1992), and Copper's very first horror tale, originally written (but never published) nearly *sixty* years earlier. He recalled: "I wrote 'The Grass' when I was an apprentice journalist aged about sixteen or seventeen in 1940, and I wrote it in my newspaper office late at night when German bombers were droning overhead and explosions sounding all around. I put it away in a drawer with a lot of other juvenilia and only dug it out about year or two ago, and had only to do a little repolishing before it was fit for publication."

Two years later I suggested that Robert Morgan's Sarob Press, one of the most impressive of all British small presses, should publish Basil Copper's next collection of "Tales of Terror and Suspense", and this appeared under the title *Cold Hand on My Shoulder* in March 2002, with illustrations and a stunning wrap-around jacket picture by Randy Broecker, limited to 300 copies, plus a further fifty-two copies in a deluxe signed and lettered edition.

The title was inspired by a horrific memory of the author's wartime experience off the coast near Portland in 1943, while attempting to pick up survivors from boats sunk by German E-boats. While Copper was struggling to lift a corpse from the sea, his colleague stumbled, and the man's dead hand fell across—and around—Copper's shoulder in a ghastly embrace.

Following the thirty macabre tales contained Copper's first six collections, and the eleven in *Whispers in the Night*, the nine stories in *Cold Hand On My Shoulder* brought the total number to exactly fifty.

A further eight "Tales of Twilight and Torment" (including six previously unpublished) were collected under the title *Knife in the Back*, privately printed by Cauchemar in 2005, limited to only 150 copies signed by the author.

Following the first UK edition of *The House of the Wolf* in 2003, Robert Morgan at Sarob Press was determined to issue the remainder of Copper's unpublished Pontine canon, beginning with his novel, *Solar Pons versus the Devil's Claw* (2004), which had been only half-finished in 1979. Copper recalled: "I managed to piece it together in 1996 from original plot notes rediscovered in my files, and successfully completed the whole novel seventeen years after it was begun!"

Solar Pons: The Final Cases (2005) contained a further five adventures (all written in 1977), plus an extra case featuring Sherlock Holmes, 'The Adventure of the Persecuted Painter' (which first appeared in Mike Ashley's anthology *The Mammoth Book of New Sherlock Holmes Adventures* in 1997).

All of Basil Copper's memorable short stories and novels are enhanced by his careful research and period detail. The first requisite for an author of fantasy is imagination, but the second is meticulous craftsmanship. If you are to take your readers into new worlds, then you have to make absolutely sure that they believe in them.

Whether it be the Los Angeles of the Mike Faraday books, the "old" London of the Solar Pons mysteries, or the fog-shrouded wastelands of his Gothic stories, the terrain of Basil Copper's fiction is always utterly convincing—even though there are times when we might wish that it wasn't!

BASIL COPPER: A WORKING BIBLIOGRAPHY (with Commentary by the Author)

FOLLOWING THREE YEARS of extensive research, every effort has been made to physically inspect the items listed on the following pages. In the summer of 2007, Basil and Annie Copper were gracious enough to welcome me into their home and their lives over a number of visits. Basil gave me unprecedented and unrestricted access to his books and papers, and the information that I gathered was used to prepare the following Working Bibliography. However, inevitably, even the author himself does not possess some copies of his own material. Wherever possible, I have contacted the people involved in these missing titles for more information. However, those items that I have not seen are marked with an asterisk (*). Basil's comments are given in italics, while any additional parenthetic notes by me are contained in the square brackets. As with any bibliography, this is (and will remain) a work in progress. If anyone has any updates or corrections, I would be grateful to receive them c/o the publisher. Thank you.

by Stephen Jones

A. MACABRE & SUPERNATURAL NOVELS

Includes an alphabetical listing of novels written by Basil Copper

"I was an inveterate scribbler even as a young child and wrote dozens of stories, even novels, essays and other pieces 'for the drawer'. Ever since childhood my mind had been entrapped by the fantastic, the macabre and the terrible, and along with the imaginings of such writers as Wells, Bierce and Blackwood, I was soon ensnared by the American pulp magazines which, in the 1930s, had invaded Britain in a big way, and the gaudy yellow, purple and scarlet covers were stacked carelessly in large mounds in Woolworths and in racks at railway bookstalls.

"Stories like 'The Dunwich Horror' inspired me, as a very young apprentice journalist to type my own tales in the newspaper offices in which I worked, very late at night, after I had finished my journalistic work. As many of these pieces were composed amid the turmoil of war, with German bombers droning overhead and the distant–and not so distant–crash of explosions, the world of horror and fantasy formed an effective means of escape from the all too real horrors of World War II as experienced by the captive civilian population of Britain in those dark years." BC.

A1 THE BLACK DEATH

Illustrations by Stefanie Kate Hawks.

a. __________, Fedogan & Bremer, Minneapolis, Minnesota, 1991 (hc)

There are two variants of this edition.

i. Description: 2,500-copy trade edition

Price: $32.00 / 373pp / Cover art by Stefanie K. Hawks.

ii. Description: 100-copy limited edition in slipcase

Price: $60.00 / 373pp / Cover art by Stefanie K. Hawks.

("The Black Death *is the longest and perhaps most ambitious Gothic novel I have ever written, with a wealth of detail. Set in the period 1895-1900, it runs to over 400 pages, has a strong romantic thread running through it, as all the best novels of this type must have, and the milieu is a village and great house in the bleak wilds of Dartmoor. In some ways it is my own* homage *to Sheridan Le Fanu, another of my favourite writers, but of course it is in no other way derivative."* BC.)

A2 THE CURSE OF THE FLEERS

a. __________, Harwood-Smart Publishing, Blandford, Dorset, 1976 (hc)
Price: £3.50 / 140pp / Cover art by Dave Austin.
b. __________, St. Martin's Press, New York, New York, 1977 (hc)
Price: $7.95 / 140pp / Jacket by Trudy Rosen.
[Note: An unpublished 225pp manuscript exists of the complete version of this novel.]

A3 THE GREAT WHITE SPACE

a. __________, Robert Hale & Company, London, England, 1974 (hc)
Price: £1.90 / 192pp / Jacket photograph by Colin Andrews.
b. __________, St. Martin's Press, New York, New York, 1975 (hc)
Price: $6.95 / 192pp / Jacket photograph by Colin Andrews.
c. __________, Manor Books, Inc., New York, New York, 1976 (pb)
Price: $1.25 / 192pp.
d. __________, Sphere Books Limited, London, England, 1980 (pb)
Price: £1.10 / 188pp.
e. [as **DIE EISHÖLLE**], Festa, Almersbach, Germany, February 2002 (hc)
Price: €17.00 / 176pp.
[Note: #8 in "H.P. Lovecrafts Bibliothek des Schreckens" series.]

A4 THE HOUSE OF THE WOLF

Illustrations by Stephen E. Fabian.

a. __________, Arkham House Publishers, Sauk City, Wisconsin, 1983 (hc)
Price: $14.95 / 298pp / Cover art by Stephen E. Fabian.

("*My only novel which deals with lycanthropy,* The House of the Wolf, *was published by Arkham House in 1983 in a beautiful edition sporting a superb colour dust-jacket and over forty brilliant line-drawings heading each chapter by the gifted American book illustrator, Stephen E. Fabian.*

"*The book took me many months as its progress was interspersed with the creation of my detective novels featuring Mike Faraday, together with my short stories and novels in the macabre field. [I sometimes wrote eight books a year when I was really in my stride.]*" BC.)

b. __________, Sarob Press, Kidwelly, Wales, August 2003
Introduction by Basil Copper. Afterword: Keeping the Wolf at the Door by Stephen Jones. Illustrations by Stephen Jones.
There are three variants of this edition.

i. Description: 150-copy limited edition (hc).
Price: £25.00; $42.50 / 207pp / Cover art by Randy Broecker.

ii. Description: 52-copy deluxe signed and lettered edition (hc).
Price: £48.00; $80.00 / 207pp / Cover art by Randy Broecker.

iii. Description: paperback edition (pb).
Price: £10.95; $18.95 / 207pp / Cover art by Randy Broecker.

(*"All I can say about this new edition is that it is fit to stand beside its US counterpart."* BC.)

A5 INTO THE SILENCE

a. __________, Sphere Books, London, England, 1983 (pb)
Price: £1.75 / 243pp / Cover art by Les Edwards.
(*"My other favourite novel, this time in the fantasy field, is* Into the Silence, *set in the 1920s, with a wealth of period detail, about an American-led expedition drilling into the bowels of the Earth in Cornwall. This was a fitting companion piece to* The Great White Space, *but it appeared briefly in paperback and then disappeared. Perhaps that too will resurface, both literally and metaphorically, in time to come."* BC.)

A6 NECROPOLIS

Author's Note. Illustrations by Stephen E. Fabian.

a. __________, Arkham House Publishers, Sauk City, Wisconsin, 1980 (hc)
Price: $12.95 / 352pp / Cover art by Stephen E. Fabian.

b. __________, Sphere Books Limited, London, England, 1981 (pb)
Price: £1.50 / 372pp.

(*"The idea of a funeral train carrying corpses and mourners of all denominations and nationalities from the heart of London on a railway line to the specially constructed Brookwood Cemetery in Surrey, is something that could probably have only evolved in Victorian England, where a great, gloomy ritual surrounded the fact of death."* BC.)

REMEMBERING AUGUST DERLETH & ARKHAM HOUSE

AS I HAVE WRITTEN ELSEWHERE, I first came across the name of August Derleth in the musty pages of *Weird Tales*, back in the 1930s when they were stacked in great mounds in Woolworths, their brilliantly-lurid colour jackets arresting the attention of passers-by.

I was only a youngster, but I have never forgotten the thrill of handling them. Memory plays one tricks, but the stories might have included 'Bat's Belfry' and 'Mr. George', one of his most delicate tales (as by "Stephen Grendon").

Years later, I was inspired to write to August when I was a newspaper editor and just launching my literary career. He replied in a jocular and friendly fashion.

August had already heard of me and was familiar with some of my work in the macabre field. He published two of my tales, 'The House by the Tarn' and 'The Knocker at the Portico' in *Dark Things* (1971), a beautifully-produced hardcover, like all his volumes.

He was the complete bookman, a scholar, a giant in physique, a man with an enormous appetite for life. He had a wonderful sense of humour, and when people rang him he would respond in a sepulchral voice, "Arrrkhaaam Hooouuussseee!"

So it was a dreadful shock to his countless friends, authors and many in a wider circle when he suddenly died after surgery and a massive heart attack.

August Derleth was succeeded by James Turner, a brilliant scholar and a meticulous craftsman so far as the minutiae of book production was concerned.

by Basil Copper

He and the lawyer who then ran Arkham House were extremely enthusiastic for my work and I was privileged to know them and August, of course.

James died tragically young of a brain tumour, and August was by no means an old man. God bless them both.

I have a great cache of correspondence with August and James in my study, which will eventually go to my literary archive at Boston University in the care of the Archivist, Professor Howard Gotlieb, so future generations will be able to study the delightful thoughts of two great bookmen.

— BASIL COPPER
Sevenoaks, England
July 2005

ROYALTY STATEMENT

ARKHAM HOUSE

PUBLISHERS

SAUK CITY **WISCONSIN**

DATE 28 February 1978

AUTHOR Basil Copper

TITLE FROM EVIL'S PILLOW
AND AFTERWARD, THE DARK

ADDRESS

AGENT

ROYALTIES EARNED ON COPIES SOLD FROM 7/1 thru 12/31/77

FROM EVIL'S PILLOW

63 copies sold ... @ 60¢		$ 37.80	royalties earned
LESS: purchases			
invoice dated 11/23/77 (air postage new book)	$ 3.05		
invoice dated 12/13/77 (thru Jim)	28.44	-31.49	
		$ 6.31	PAYMENT HEREWITH

AND AFTERWARD, THE DARK

Total advances against royalties	$1750.00	
560 copies sold ... @ 75¢	- 420.00	royalties earned
	$1330.00	Unearned balance as of 12/31/77

B. MACABRE & SUPERNATURAL COLLECTIONS

Includes an alphabetical listing of collections containing short stories and novellas by Basil Copper

(*"For me, the horror story pure and simple–though it is neither pure nor simple–must be meticulously worked out as to specifics before the fingers touch the typewriter keys; not the detail, of course. The colour comes as one proceeds; but the denouement, the basic idea must come to mind first. The atmosphere follows but the plot, whether it be simple or convoluted, must lead unerringly to that shattering finale, preferably in the last two or three sentences."* BC.)

B1 AND AFTERWARD, THE DARK: SEVEN TALES

Contents: Introduction by Edward Wagenknecht, *The Spider, The Cave, Dust to Dust, Camera Obscura, The Janissaries of Emilion, Archives of the Dead, The Flabby Men.*

a. _________, Arkham House, Sauk City, Wisconsin, 1977 (hc).
Price: $7.50 / 222pp / Cover art by Stephen E. Fabian.
[Note: Published in an edition of 4,000 copies.]

B2 COLD HAND ON MY SHOULDER: TALES OF TERROR & SUSPENSE

Contents: Introduction by Richard Dalby, *Death of a Demi-God, As the Crow Flies, Poetic Justice, Bright Blades Gleaming, When Greek Meets Greek, Ill Met by Daylight, Charing Cross-Dover-Charing Cross, Line Engaged, Beyond the Reef.* Illustrated by Randy Broecker.

a. _________, Sarob Press, Mountain Ash, Wales, March 2002 (hc)
There are two variants of this edition.

i. Description: 300-copy limited edition.
Price: £25.00; $45.00 / 199pp / Cover art by Randy Broecker.

ii. Description: 52-copy signed and lettered deluxe edition.
Price: £50.00; $85.00 / 199pp / Cover art by Randy Broecker.

B3 FROM EVIL'S PILLOW

Contents: *Amber Print, The Grey House, The Gossips, A Very Pleasant Fellow, Charon.*

a. __________, Arkham House, Sauk City, Wisconsin, 1973 (hc)
Price: $6.00 / 177pp / Cover art by Frank Utpatel
[Note: Published in an edition of 3,500 copies.]

B4 HERE BE DAEMONS: TALES OF HORROR AND THE UNEASY

Contents: *Old Mrs Cartwright, The Knocker at the Portico, The Way the World Died, The Second Passenger, The Treasure of Our Lady, Justice at the Crossroads, Mrs Van Donk, The Trodes, The Great Vore.*

a. __________, Robert Hale Limited, London, England, November 1978 (hc)
Price: £4.50 / 219pp / Jacket design by Colin Andrews.
[NOTE: A 1978 uncorrected proof copy of this title exists.]

b. __________, St. Martin's Press, New York, New York, 1978 (hc)
Price: $8.95 / 219pp / Jacket design by Colin Andrews.

c. __________, Sphere Books, London, England, 1981 (pb)
Price: £1.25 / 213pp.

B5 KNIFE IN THE BACK: TALES OF TWILIGHT AND TORMENT

Contents: *There Lies the Danger . . ., Death of a Nobody, Reflections, The White Train, Hunted by Wolves, The Candle in the Skull, Storm Over Stromjolly, The Silver Salamander.*

a. __________, Cauchemar [Sarob Press], Sevenoaks, 2005 (hc)
Privately printed / 71pp.
[Note: Published by Sarob Press on behalf of the author in an edition of 150 signed and numbered copies.]

B6 NOT AFTER NIGHTFALL: STORIES OF THE STRANGE AND THE TERRIBLE

Contents: Foreword, *The Spider, Camera Obscura, The Cave, The Grey House, Old Mrs Cartwright, Charon, The Great Vore, The Janissaries of Emilion.*

a. __________, New English Library/Four Square, London, England, June 1967 (pb)
Price: 5/- / 190pp / Cover photograph by Michael Busselle.

B7 WHEN FOOTSTEPS ECHO: TALES OF TERROR AND THE UNKNOWN

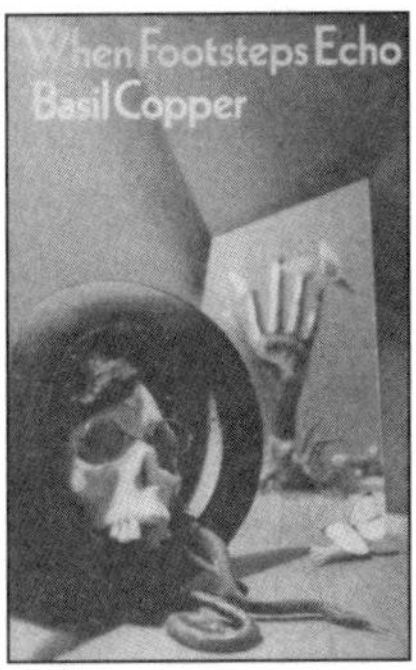

Contents: *Camera Obscura, The Janissaries of Emilion, Amber Print, A Message from the Stars, Doctor Porthos, Cry Wolf, The Academy of Pain, The Recompensing of Albano Pizar, Out of the Fog, Archives of the Dead.*

a. __________, Robert Hale & Company, London, England, 1975 (hc)
Price: £2.25 / 184pp.
b. __________, St. Martin's Press, New York, New York, 1975 (hc) 184pp.

B8 VOICES OF DOOM: TALES OF TERROR AND THE UNCANNY

Contents: *The House by the Tarn, The Stranger, Charon, A Very Pleasant Fellow, The Gossips, The Madonna of the Four-Ale Bar, The Grey House.*

a. __________, Robert Hale Limited, London, England, March 1980 (hc)
Price: £5.25 / 190pp / Jacket photography by Colin Andrews.
[Note: A 1980 uncorrected book proof exists of this edition.]
b. __________, St. Martin's Press, New York, New York, 1980 (hc) 190pp / Jacket photography by Colin Andrews.

B9 WHISPERS IN THE NIGHT: STORIES OF THE MYSTERIOUS AND MACABRE

Contents: Introduction: Out of the Fog: Recollection by Stephen Jones, *Better Dead, Reader, I Buried Him!, One for the Pot, Wish You Were Here, In a Darkling Wood, The Grass, Riding the Chariot, Final Destination, The Obelisk, Out There, The Summerhouse.*
Illustrated by Stephen E. Fabian.

a. __________, Fedogan & Bremer, Minneapolis, Minnesota, March 1999 (hc)

There are two variants of this edition.

i. Description: 1,000-copy trade edition.
Price: \$27.00 / 271pp / Cover art by Stephen E. Fabian.

ii. Description: 100-copy limited edition in slipcase.
Price: \$95.00 / 271pp / Cover art by Stephen E. Fabian.

ARKHAM HOUSE PUBLISHERS, Inc.
SAUK CITY, WISCONSIN 53583
(608) 643-4500

4 June 1977

Mr Basil Copper
England

Dear Basil,

The <u>And Afterward, the Dark</u> galleys finally straggled in late Friday afternoon, but I still had time to rush your set to the local post office that same day. I have glanced at the texts briefly, and our compositor appears to have been on fairly good behavior, but I haven't had the opportunity to READ these yet, either!

Please correct these carefully and return them to me at your earliest possible convenience.

Best wishes,

JT

THE GAME'S AFOOT! AUGUST DERLETH & SOLAR PONS

IT WAS IN THE EARLY 1920S when a young student at the University of Wisconsin, who was interested in becoming an author, wrote to the great Sir Arthur Conan Doyle in England and asked if he intended to write any more Sherlock Holmes stories. Sir Arthur replied that he did not. The young man then wrote back to ask if the famous writer would mind if he himself wrote some more Holmes tales. The answer was in a decided affirmative! Unfazed, the young man sat down to write his own Sherlockian pastiches. He called his detective hero Solar Pons–he once explained the derivation [of the name] to me as "Bridge of Light"–and his Watson counterpart Dr. Lyndon Parker.

The young man was, of course, August Derleth, and in those far-off days in rural Wisconsin he could not have foreseen how celebrated he was to become. He eventually got some of the stories published in 1929 and later became famous in America in an amazing diversity of roles.

For he was a polymath; a giant in stature, in energy and in the esoteric world he had chosen to call his own. A man of enormous appetite for life, for the good things like food and wine, and for his literary labours; he took on a workload that endured decade after decade, and that would have crippled any ordinary human being and which, tragically, finally ended his life at the age of 62, after heart complications.

August was a poet, an essayist, founder of the world's most celebrated macabre and fantasy publishers, Arkham House, with the late Donald

by Basil Copper

Wandrei—and its associated imprint, Mycroft and Moran; a Hollywood scriptwriter; newspaper columnist; author of hundreds of books, many of them published by such distinguished New York houses as Scribners, but the majority issued under his own Arkham House logo—and thousands of newspaper and magazine articles; university lecturer; public speaker, and he also recorded his own poems on discs and wrote forewords for other writers' work.

In addition to all this he had assembled perhaps the world's largest collection of comic strips; had written the "Judge Peck" novels; massive regional novels set in the historical context of his beloved Wisconsin, where he spent weeks tramping the woods in search of his great culinary delicacy, morels, a type of mushroom much favoured by connoisseurs; had written eight volumes of "Solar Pons" adventures, issued in hardback under his own imprint; books on Catholic themes which gained him the Apostolic Blessing of Pope John in 1959; and he was also awarded a Guggenheim Fellowship in 1938. August was highly gratified at being praised by so distinguished an American writer as Sinclair Lewis, which must have sustained him through many bleak years of toil in which, he once told me, he worked against sixty-day promissory notes, in order to support his family of two young children, his aged mother, and the beautiful home he had custom built, Place of Hawks, where his own library and Arkham book stocks were kept.

He was a passionate book collector, amassing a library of some 12,000 volumes; a chess enthusiast; a strong swimmer; a writer of children's books; a devoted father and family man; oversaw films of his tales made for TV; and he had written hundreds of macabre short stories and novellas, many of which have become classics in the genre. As if all this were not enough, he edited volumes of ghost and macabre stories, which have appeared in both hardback and paperback on a world scale, and almost single-handedly thrust the great horror writer H.P. Lovecraft on to the world stage with his patronage and tireless promotion of the former's work through publishing all of his novels and short stories under the Arkham imprint. And with this list I have by no means exhausted all the range and scope of his activities in the literary world. A Renaissance man, then, as I have often described him, both in print and in lectures.

My own association with August Derleth began many years ago when, as a newspaper editor, I was at the same time embarking on a career as a professional author of novels, short stories and non-fiction works. I had written to him to see if there was any possibility of his publishing some of my macabre tales and he replied most warmly. He had already seen some of my horror stories in the famous Pan Books series and, as I have recorded elsewhere, his encouragement, long and enjoyable letters and his patronage when he offered to publish my first American hardback under the aegis of Arkham House, formed one of the warmest and most delightful strands of my writing life. Regretfully, our relationship, conducted entirely on paper, was to last only some five or six years, but during that time August emerged as a man of great probity, courage and generosity.

His death in 1971 at the early age of 62 was a staggering blow to his host of friends and literary colleagues, and a world-wide loss in the publishing field. A brilliant editor, James Turner, then took over the running of Arkham House, and he published a number of my horror story collections, as well as major Gothic novels of mine. But one day I wrote to him after he had sent me some copies of August's Pons volumes and I pointed out a number of errors in the texts. There is a story behind this which I will come to a little later. The upshot was that Arkham then commissioned me to revise and edit the entire Pontine Canon—some eight volumes—eventually published by Arkham in a magnificent two-volume slip-cased edition as *The Solar Pons Omnibus* in 1982.

This definitive work collected together everything August had ever written about Pons and the contents of the two volumes, running to over 1,300 pages, with fine illustrations by the late Frank Utpatel and with a foreword by Robert Bloch, brought together: *The Casebook of Solar Pons*; *The Chronicles of Solar Pons*; *"In re Sherlock Holmes"–The Adventures of Solar Pons*; *The Memoirs of Solar Pons*; *The Reminiscences of Solar Pons*; *The Return of Solar Pons*; the novel *Mr. Fairlie's Final Journey*; and A *Praed Street Dossier.* A veritable feast for Pontine enthusiasts, bringing together as it did some seventy short stories and a novel, together with the fine and appreciative tribute by Robert Bloch. Many of the stories achieved much wider circulation as the years went by when they were anthologised in hardback and paperback from distinguished imprints in both the English-speaking world and in Continental Europe.

The editing of the Pons Canon was carried out as a labour of love and was a most enjoyable task, which occupied me for almost a year, in between other writing assignments. It was undertaken as a tribute to August's memory and due to my affection for the man, and I did not wish any payment for the work, but Arkham eventually insisted that I accept a fee, which was entirely in keeping with their scrupulous business ethics. It was some while after this that James Turner asked if I would like to carry on the Pons series myself. Of course, I accepted immediately, and over the next two years, I wrote some six volumes. These were *The Dossier of Solar Pons*; *The Secret Files of Solar Pons*; *The Exploits of Solar Pons*; *The Further Adventures of Solar Pons*; *Some Uncollected Cases of Solar Pons*, and *The Recollections of Solar Pons,* together with an almost completed novel, *The Devil's Claw*. All the stories in the collections were of novella length, some running to over seventy pages, which gave me much greater scope for character, plot development and all those macabre elements of atmosphere which are so dear to my heart.

They were received with great enthusiasm by James Turner, but unfortunately they had a chequered history as paperbacks over a number of years, by imprints which completely destroyed the 1920s atmosphere by rewriting and inserting crass and stupid alterations. One went bankrupt, and others did not pay or ever issue royalty statements. Fortunately, I now have a splendid new publisher in the U.S.A., Fedogan & Bremer, who in addition to producing a magnificent edition of my Gothic novel, *The Black Death*, have now started issuing the unpublished Solar Pons tales, the first volume of which, *The Exploits of Solar Pons*, was launched with great success when I was Guest of Honour at the World Fantasy Convention in Minneapolis in October/November 1993.

Both these books were superbly illustrated by my close friend and collaborator, the American artist, Stefanie Kate Hawks, and the publishers sold hundreds of copies at the Convention, when they organised signing sessions for me which lasted for hours. One of the most successful was at the Sherlock's Home pub in Minnetonka, when a huge crowd of Solar Pons enthusiasts, who included distinguished doctors, surgeons and anaesthetists turned up. A beautiful girl cabaret artiste played piano; a genuine Scottish piper piped me in; and a well-known local radio actor complete with deerstalker announced me in stentorian tones!

In my speech I said that one of the most notable absentees at the function was the creator of Solar Pons, August Derleth, and I underlined my own association with the series, and the beautiful volume of *The Exploits*, produced by Philip Rahman and Dennis Weiler, respectively computer expert and surgeon, under the personas of Fedogan & Bremer, again sold in astonishing numbers. While paying tribute to August I outlined some of the difficulties he had when publishing his own series of Pons tales. I said he had been a great friend to me and to many others; a gifted writer who wore himself out in the service of literature and in running his imprint, Arkham House. A giant in so many ways, not least in stature, and a person of absolute probity and honesty in all his dealings, often paying in advance for work that would not see print for many years.

A long time ago August asked the advice of a well-known British author and commissioned him to go through his entire Pons canon and make the necessary corrections and give advice on British usage, etc. The man, who is now dead, and was a noted Sherlockian, treated August very badly. When I edited the two-volume edition in 1981, working from the original printed texts, I found that little or no work had been done to insert the correct terms, and I understood that August had paid him thousands of dollars for the "work". I found some 2,000 errors of British usage when engaged on my rescission of the texts, I told my audience. For instance, when describing British railway procedures, August would say, "The cars were switched to the loop-line", whereas the correct British terminology would be, "The carriages were shunted on to a siding". On another occasion the author had a train from the North of Scotland speeding straight to Dover without stopping. Of course, most of the major London stations are termini–notable exceptions being London Bridge and Waterloo East–so trains from the north would terminate at such destinations as Paddington and King's Cross. Passengers would then travel by Underground or taxi to Charing Cross and an onward train to Dover. August was not to blame for this as he well knew that there was something wrong with the usage–he had in fact never visited Britain–but it was shameful that the author employed did not do the job. But those many thousands who now read the definitive texts will find it very hard indeed to detect even one fault of usage or terminology. It was the least I could do in tribute to August in undertaking the work.

What of the stories themselves? Some are more successful than others, of course. Pons had first appeared in 1929 and August was still at work on *The Chronicles of Solar Pons* at the time of his death in 1971. In fact, it was probably the last completed work that he wrote for the material was on his desk at his spacious home, Place of Hawks, waiting to be re-typed. Of course, the characters had evolved more as the years had passed, though the Pontine characteristics–which I have closely emulated–were still to the fore; notably his habit of "tenting" his thin fingers before him; or pulling at the lobe of his ear. Little habits which had helped to endear him to generations of readers over the more than forty years of his appearances, during the Derleth tenure of the familiar sitting room in Praed Street.

The Chronicles was published in 1973 under the Mycroft and Moran imprint with an introduction by Allen J. Hubin, and contains one of the original Canon's most published stories which has appeared frequently in anthologies on both sides of the Atlantic, in the intervening twenty-one years, 'The Adventure of the Benin Bronze'. It begins intriguingly with a cryptic drawing of three globular objects and an eye sent to Randall Creighton, a Basingstoke man who had been in the British Army in Africa and had taken part in a horrific battle there. An ingenious plot involving murder, the theft of priceless bronzes and a cunning red herring are some of the ingredients of an absorbing tale.

There are other favourites, all of which are remarkable when viewed against Derleth's entire output and the fact that he wrote many of his tales in rooms where his students were drinking and talking at night, during his stints as a university lecturer. In this he paralleled Georges Simenon who worked away at his typewriter, while running noisy drinks parties in his rooms in Paris during the 1920s. 'The Aluminium Crutch' is another outstanding tale and there are many others while readers world-wide will have their favourites. It has been a particular pleasure to me to carry on these fascinating adventures where August left off and judging by both critical reaction and by word of mouth, the reception for my own first hardback volume of previously unpublished Pons novellas, means that the adventures of the celebrated Praed Street detective and his loyal colleague, Dr. Lyndon Parker, will go on for many years to come.

One of the great pleasures of being a writer and one also greatly enjoyed by August Derleth is the feedback from readers, some of them distinguished fellow authors; this may be years in coming–it is very often like dropping a pebble down a deep well and waiting for the muted splash–but when it does it is an undiluted joy, especially if it be praise for the author or pleasure experienced by the reader. The reception for my own *Exploits of Solar Pons* in Minneapolis was overwhelming and similar reactions were forthcoming at the necessarily much smaller World Horror Convention in Phoenix, Arizona, in March 1994, which I attended.

For the writer's life is essentially a lonely one, facing the wall while toiling at the typewriter day after day, year after year, and it is pure joy to meet readers, young and old, face to face and get their reactions. This August also experienced in great measure and I am sure it gave him equal pleasure. My tributes to his memory with my own Pontine volumes will be continuing over the coming years and together with his own works in the genre will help to keep a great human being's memory green. August has gone but his work endures.

C. THE "SOLAR PONS" SERIES

Includes an alphabetical listing of novels and collections written by Basil Copper, based on the characters created by August Derleth

(*"I am not writing* new *Solar Pons stories. My Solar Pons material was authorised and commissioned by Arkham House Publishers, Inc., back in the 1970s, and was written then but, owing to changes at Arkham House, they were never published by that imprint. The tales remained in limbo for something like twenty years before the originals began to be published by Fedogan & Bremer of Minneapolis. All date from the 1970s and early 1980s, so perhaps original copyright lines should have been printed together with the new publication dates, which would have obviated all misunderstandings. So, all the Pons tales from my pen were commissioned and authorised by the then editorial and legal personnel at Arkham House (who, incidentally, were hoping that I would write the twelve volumes I had originally promised them) and by the estate of the late August Derleth, who was a great friend of mine and very enthusiastic about my work."* BC.)

C1 THE ADVENTURE OF THE SINGULAR SANDWICH

"Special Promotional Edition". Contents: *The Adventure of the Singular Sandwich,* A Conversation with Basil Copper conducted by R. Dixon Smith, A Brief Note on Solar Pons by the Publishers. Illustrations by Stefanie K. Hawks.

a. __________, Fedogan & Bremer, Minneapolis, Minnesota, 1995 (chapbook).
50pp.
[Note: "A Solar Pons Sampler"].

C2 SOLAR PONS #8: THE DOSSIER OF SOLAR PONS

Short story collection. Contains: *The Adventures of the Perplexed Photographer, The Sealed Spire Mystery, The Adventures of the Six Gold Doubloons, The Adventure of the Ipi Idol, The Adventure of Buffington Old Grange, The Adventure of the Hammer of Hate.*

a. _________, Pinnacle Books, Los Angeles, California, January 1979 (pb)
Price: $1.95 / 310pp / Cover art by Ben Stahl.
[Note: "Based on the characters and series created by August Derleth." This was originally intended to be an Arkham House book.]
b. [as **THE DOSSIER OF SOLAR PONS**], Academy Chicago Publishers, Chicago, Illinois, 1987 (pb)
Price: $5.95 / 278pp / Cover design and illustration Mark Coleman.

C3 THE EXPLOITS OF SOLAR PONS

Short story collection. Contains: *The Adventure of the Verger's Thumb, The Adventure of the Phantom Face, Death at the Metropole, The Adventure of the Callous Colonel.* Illustrated by Stefanie K. Hawks.

a. _________, Fedogan & Bremer, Minneapolis, Minnesota, 1993 (hc)
There are two variants of this edition.
i. Description: 1,900-copy trade edition
Price: $24.00 / 239pp / Cover art by Stefanie K. Hawks.
ii. Description: 100-copy limited edition in slipcase
Price: $60.00 / 239pp / Cover art by Stefanie K. Hawks.

(*"Horror and the Gothic are heavily overlaid in my Solar Pons series though the supposed supernatural happenings are often found to be caused by malign but human agencies. This is an old tradition, of course, as exemplified in Conan Doyle's* The Hound of the Baskervilles, *and it is one of which I am very fond."* BC.)

C4 SOLAR PONS #9: THE FURTHER ADVENTURES OF SOLAR PONS

Short story collection. Contains: *The Adventure of the Shaft of Death, The Adventure of the Defeated Doctor, The Adventure of the Surrey Sadist, The Adventure of the Missing Student.*

a. _________, Pinnacle Books, Los Angeles, California, August 1979 (pb)
Price: $1.95 / 208pp / Cover illustration by Ben Stahl.
[Note: "Based on the characters and series created by August Derleth." This also includes a four-page essay, 'Introducing Doctor Who', by Harlan Ellison.]

C5 THE FURTHER ADVENTURES OF SOLAR PONS

Short story collection. Contains: *The Adventure of the Shaft of Death, The Adventure of the Frightened Governess, The Adventure of the Defeated Doctor, Murder at the Zoo.*

a. __________, Academy Chicago Publishers, Chicago, Illinois, 1987 (pb)
Price: $5.95 / 250pp / Cover design and illustration by Mark Coleman.

C6 THE RECOLLECTIONS OF SOLAR PONS

Short story collection. Contains: *The Adventure of the Mad Millionaire, The Adventure of the Cursed Curator, The Adventure of the Hound of Hell, The Adventure of the Singular Sandwich.* Illustrated by Stefanie K. Hawks.

a. __________, Fedogan & Bremer, Minneapolis, Minnesota, 1995 (hc)
There are two variants of this edition.
i. Description: 1,900-copy trade edition
Price: $25.00 / 250pp / Cover art by Stefanie K. Hawks.
ii. Description: 100-copy limited edition in slipcase
Price: $75.00 / 250pp / Cover art by Stefanie K. Hawks.
[Note: "The novella 'The Adventure of the Singular Sandwich', which briefly appeared in paperback in a much bowdlerised and altered form, appears here for the first time with the definitive text."]

C7 SOLAR PONS #10: THE SECRET FILES OF SOLAR PONS

Short story collection. Contains: *The Adventure of the Crawling Horror, The Adventure of the Anguished Actor, The Adventure of the Ignored Idols, The Adventure of the Horrified Heiress.*

a. __________, Pinnacle Books, Los Angeles, California, October 1979 (pb)
Price: $1.95 / 210pp / Cover illustration by Ben Stahl.
[Note: "Based on the characters and series created by August Derleth."]

C8 THE SOLAR PONS OMNIBUS

by August Derleth.

Short story collection. Drawings by Frank Utpatel.

a. __________, Volume One, Arkham House Publishers/Mycroft & Moran, Sauk City, Wisconsin, 1982 (hc)
Foreword by Robert Bloch.

b. _________, Volume Two, Arkham House Publishers/Mycroft & Moran, Sauk City, Wisconsin, 1982 (hc)
Price: $39.95 / 1,306pp.
[Note: Edited and corrected by Basil Copper, this was published in a slipcased two-volume set.]

C9 **SOLAR PONS THE FINAL CASES**

Short story collection. Contains: Foreword by the author, *The Adventure of the Haunted Rectory, The Adventure of the Ignored Idols, The Adventure of the Horrified Heiress, The Adventure of the Baffled Baron, The Adventure of the Anguished Actor*, & a 'Sherlock Holmes' case *The Adventure of the Persecuted Painter.*

a. _________, Sarob Press, Kidwelly, Wales, 2005 (hc)
There are two variants of this edition.
i. Description: 220-copy limited edition.
Price: £27.50; $49.50 / (vii) 183pp / Cover art by Les Edwards.
ii. Description: 55-copy signed deluxe edition in slipcase.
Price: £65.00; $115.00 / (vii) 183pp / Cover art by Les Edwards.

(*"It should be noted that the entire canon of my Solar Pons stories following the series written by the late August Derleth was commissioned and authorised by Arkham House USA back in the mid-1970s but, for various legal reasons, were withheld from publication and are only now appearing in their original form after some thirty years in limbo. Over the decades a number of editions with altered and mutilated texts have appeared."* BC.)

C10 **SOLAR PONS VERSUS THE DEVIL'S CLAW**

Novel. Author's Note. Introduction by Richard Lancelyn Green.

a. _________, Sarob Press, Kidwelly, Wales, 2004 (hc)
There are two variants of this edition.
i. Description: 200-copy limited edition.
Price: £25.00; $45.00 / (xiii) 164pp / Cover art by Les Edwards.
ii. Description: 50-copy signed and numbered deluxe edition in slipcase.
Price: £55.00; $95.00 / (xiii) 164pp / Cover art by Les Edwards.

(*"Though written some twenty-five years ago, my only Solar Pons novel has never before been published, so this is its first time in print."* BC.)

C11 SOLAR PONS #11: THE UNCOLLECTED CASES OF SOLAR PONS

Short story collection. Contains: *The Adventure of the Haunted Rectory, The Adventure of the Singular Sandwich, Murder at the Zoo, The Adventure of the Frightened Governess.*

a. __________, Pinnacle Books, Los Angeles, California, January 1980 (pb)

Price: $1.95 / 209pp / Cover illustration by Ben Stahl.

[Note: "Based on the characters and series created by August Derleth." The Author's original title was *Some Uncollected Cases of Solar Pons.*]

LES EDWARDS

AUGUST DERLETH: A GIANT REMEMBERED

IT BEGAN IN THE LATE 1920s and early '30s when I, a very young schoolboy, first became aware of the harsh, pungent aroma of American pulp-fiction. These publications have long since disappeared from British bookstalls, but I can close my eyes and smell them now, plunge again in memory into the already yellowing pages.

For the American pulp magazine had invaded Britain in a big way and the gaudy yellow, purple and scarlet covers were stacked carelessly in large mounds in Woolworth's and in racks at railway bookstalls. The cowboy and gangster magazines, notably *Black Mask*, which contained the early short stories of Chandler, interested me not at all.

My mind had already been entrapped by the fantastic, the macabre and the terrible, and along with the imaginings of such writers as Wells, Bierce and Blackwood, I was immediately taken by the statuesque nudes and Laocoon-like writhings of squamous, blasphemous creatures from the lower depths which enlivened or perhaps disfigured—it is impossible to make judgement now at this seemingly vast distance in time—the covers of *Weird Tales*.

Amazing to recall that they were available to every comer then at the counter of my particular Woolworth's in Kent for an incredible threepence apiece and I had dozens of them in my pulp fiction collection. All gone with the wind, alas, and not one single copy remains to romanticise my shelves.

At about the same time I began desultorily collecting the famous "Not at

by Basil Copper

Night" series of horror stories, edited by Christine Campbell Thomson and published by Selwyn & Blount at the amazingly low price of two shillings. They had first appeared in 1925, but it was not until much later that I discovered them, probably around 1935 when their price had not risen by as much as a penny.

If I remembered correctly there were twelve volumes in the original series, they sold in stupefying quantities—tens of thousands—and these volumes, together with the *Weird Tales* editions, first introduced me to some writers who were later to make giant reputations. Strangely enough, I recall only a few titles from this period.

For by now the macabre had taken firm hold and I devoured every book, classic or otherwise, that I could get hold of between the ages of ten and eighteen. Slowly, favourites began to emerge; I remember once, at about the age of twelve, being terrified by a story in *Weird Tales* which concerned a thing which was disembowelling sheep on a lonely hillside.

I rediscovered the piece, graced by good paper, decent type and between hard-covers many years later; it could only have been H.P. Lovecraft's celebrated 'The Dunwich Horror' which chronicled the terrible adventures of the immortal Wilbur Whateley. Names were emerging now, both in the pages of *Weird Tales* and in the "Not at Night" series.

One of them had written some stories which took my fancy. Together with Lovecraft, Henry S. Whitehead, Hazel Heald (I did not then know of Lovecraft's ghost-writing) and a handful of others, he became my favourite. Remember, that in the 1930s and even the '40s, good macabre writers were few and far between and their work had to be sought in the pages of the anthologies that were issued sparsely and at long intervals over the years, or in the occasional single-writer collections. All in hardback, for there were few paperbacks then.

Some titles remained in the mind, began to reappear as anthologies became more frequent. A particular favourite was 'Prince Borgia's Mass'. Another was 'The Tenant'. A third 'The Extra Passenger'. Their author was a men named August Derleth.

In the pulp magazines, sandwiched among advertisements for trusses and cures for acne, it was difficult to take even the best macabre fiction seriously. Given the dignity of hard-covers and graceful type these writers began to emerge as serious literary figures, even in the small, specialised niche they had chosen.

Years passed, but the name of August Derleth continued to emerge in a variety of spheres: as editor; as anthologist; as the champion of Lovecraft. I also became enamoured of an excellent writer of ghost stories, whose delicate-hued narratives had something in them of Algernon Blackwood and Lord Dunsany (whom I knew and met many times as a journalist).

The ghost tales ware written by a man named Stephen Grendon, and again it was a long time before I learned that he and August Derleth were the same person. I discovered Arkham House in the first two or three years of the war, when rare copies found their way to British shores; and when serving aboard a motor torpedo boat engaged in fighting German E-boats in the Channel during some of the most bitter winters of the war, I found off-duty consolation in the philosophy of Thoreau and his remembered tranquillity of Walden Pond.

Again, there was a connection with Derleth; for I discovered he was an essayist, a fine nature writer, a biographer of Thoreau and he himself had often walked the shores of Walden Pond. Peace came, more years went by, but still Derleth was a name in the background and gradually he began to permeate my consciousness in many different fields.

I pursued a career in journalism, became a newspaper editor, collected books, travelled and amassed a collection of historic films. When I was struggling to establish myself as a professional writer of novels, non-fiction studies and macabre stories nearly two decades ago his name again came before me on the fly-leaf of a book. He was an author and a publisher, was he not? But it was not until some time later that, on impulse, I wrote to him.

He replied in most cordial terms, asking to see my work. Some of my tales had already seen publication in Pan Books and I was astonished to learn that he had already heard of me. His encouragement, long and enjoyable letters and his patronage when he offered to publish my first American hardback under the imprint of his famous Arkham House, formed one of the warmest and most delightful strands of my writing life.

Unfortunate indeed that our correspondence was to last only five or six years, for his chats on paper, in which he was incredibly frank about his financial and other difficulties, were a constant example to me of honesty and courage in the face of great odds which bolstered my own resolution in times of discouragement and financial difficulty.

I have already paid public tribute to August on both sides of the Atlantic in my own non-fiction studies, so I would prefer to paint a more intimate picture of a good-humoured, generous and loveable human being in these random recollections. I am on record as saying he was a Renaissance man. This was literally true, and his huge appetite for literature and life kept him at his desk under an incredible workload that would have consumed lesser men, for decade after decade.

A great deal of his work was slipshod and careless, of course; what of that? He wrote too much, turning out nearly 200 books as well as thousands of pieces of journalism. Agreed. But at his best he could write beautifully, and much of his work will last. Surely his ghost story 'Mr. George' is one of the finest things of its kind in the language.

He was in turn poet; essayist; *pasticheur* of Sherlock Holmes in his famous "Solar Pons" series; broadcaster; Hollywood script-writer; journalist; newspaper columnist; macabre writer, both in the novel and the shorter form; champion of Lovecraft; publisher and founder of the world's finest macabre imprint, Arkham House; regional novelist with massive historical sagas like *Wind Over Wisconsin*; book column editor; anthologist; nature writer; great walker and tireless collector of comic strips, of which he had possibly the world's largest collection; collector of morels, a delicacy in the mushroom line in his search for which he tramped miles through his beloved Wisconsin woods.

He was humorous; incredibly generous, often paying one for work which would not see print for literally years. He was the last of the all-round *litterateurs*; the complete man of letters, as he often proclaimed himself. He had a huge zest for life; loved wisely and unwisely; was himself greatly loved. A man of enormous physique and enormous appetite, both for food and the good things of this life, his fall was like that of a giant oak.

I had only recently received a letter from him in hospital one beautiful summer morning in 1971 and was reading it at breakfast when I was called to the phone at 8:30 a.m. It was my old friend Richard Davis, and the news he imparted was like a physical blow. Certainly the small literary world which embraces the macabre and the fantastic was shaken to its foundations. Yet not one English daily or evening newspaper saw fit to chronicle his death.

Like Lovecraft, he passed almost unnoticed except for the gigantic ripples in the small, rather esoteric world he had chosen to make his own. So celebrated a writer as Sinclair Lewis had once in public proclaimed a great future for the young Wisconsin writer; perhaps he did August a disservice, for ever after he drew on the quotation and basked in the kudos it had brought him.

Understandable, perhaps, yet the rainbow he sought somehow always eluded him; year after year he toiled on, sometimes working against mountainous debts, supporting his family of two small children and his aged mother; his home; the publishing house he had founded; and still finding himself able to make the generous gestures toward friends and struggling writers who would perhaps never have found publication at all except within the covers of *The Arkham Sampler*, the small magazine issued as a corollary to the Arkham House imprint.

Yet he was not a tragic figure; he enormously enjoyed his life and the literary acclaim he found inside and outside his home state, and by any standards his was a gigantic achievement in so many fields. Better, perhaps, to have specialised more; to have written less; and published fewer poor writers. Yet he could not have done other than he did; for he was August Derleth, a proud, generous, kindly man, whom I shall always be glad to have known, even through the medium of the many lengthy and sometimes hilarious letters we exchanged.

It was true, thought, that August did care about the neglect of his more serious work as an important regional novelist, and he greatly regretted, as he often told me, that such books as *Evening in Spring*, *Restless is the River* and *Shield of the Valiant* had never been published in England or Europe. Ironically, though, this will probably prove to be only a matter of time.

The Solar Pons stories have been widely paperbacked in America and in addition to the two-volume hardback edition being issued by Arkham House–(I little thought in 1935 as a small child that, forty years later, I would spend some eight months editing and revising the entire canon for Arkham)–plans are now afoot to publish the tales in England.

I have myself so far written some four volumes of Solar Pons stories, following August's original model, and this has been one of the most pleasing tributes I could have paid to his memory and to the long and fruitful collaboration I have enjoyed with Arkham and its editor, James Turner.

Volumes of anthologies edited or presented by August are now in print on a world scale; in hardback from such distinguished English publishers as Gollancz; and in paperback from such diverse imprints as New English Library, Panther and Mayflower. His reputation can only increase and appreciate as the years go by, while Arkham House itself in its prosperous and steady continuance is a living memorial to his courage and his life-work.

Let me end by selecting a few typical extracts from his letters–(and I hope one day that perhaps Arkham might issue a selection chosen from the thirty-odd years that August was in charge).

In 1967 he wrote: "I am sorry, though, that my work in England seems to be limited to entertainment–none of my serious work has been published over there, and that is a matter of great regret because, for one thing, it *is* after all my best, and, for another, I've always been a profound Anglophile."

And in the same letter: "I suppose that 'Mr. George' is one of my better tales of the macabre. I am also rather fond of 'Mrs. Manifold', 'Lonesome Places', 'A Room in a House' and 'The Panelled Room . . . Like yourself, I admire the work of Robert Aickman and Roald Dahl. Dahl is rather more in a class with John

Collier, while Aickman seems to me more in the tradition of M.R. James and Wakefield."

On a postcard the same year: "I've been off teaching for a fortnight and now face all the accumulated mail – and that never comes in small amounts, often averaging fifty letters a day."

Again, a wistful note is struck a little later, in August, 1967: "It is gratifying to know that my paperbacks are well displayed and selling well over there. But I confess I would be more gratified at publication of such a book as *Walden West*, which is a considerably more solid creative achievement."

On a writer "drying up": "But dry periods come to us all, really. I know many writers, old and young, and these periods are as natural as anything in a creative individual's life. They are difficult for one-type writers, but of little moment in the case of a more versatile writer. They've never really troubled me, and I mention the instance in 'Lovecraft as Mentor' solely because it was so v. unusual in my experience. It is less so now, of course, since I've lived a much longer time—thirty years longer, to be exact."

On the macabre: "I suppose it is inevitable that I should be identified primarily as a writer in the domain of the macabre, though only a quarter of my work (including detective fiction) could be so classified, and I look upon that writing as rather an entertainment . . ."

On publishing: "I'm putting up a warehouse this coming year [1969], probably in the summer; that will slow up my book production a little, but no matter . . . I have a jr. novel to do directly I finish the revision of the biographical memoir—for it is that of HPL, rather than a biography, which I wouldn't undertake until all the Lovecraft letters have been brought out.

"Yes, of course, I have a master file of the Arkham House books. I suspect there must now be about 100 of them. But warehouse or not, I do expect to do more selective publishing beginning in a few years; I publish too many slow-moving books, and that leaves me perpetually running a hand-to-mouth business, with just enough money for the bills and none with which to enjoy myself; and, since I'll turn sixty 2/24 [1969] I do feel I'd better set up a savings account, instead of just a checking account, so that I won't be caught disastrously short in the event of an economic recession of major proportions, which, in these inflationary times, is not an improbability, no matter how much our respective governments fancy they've hedged against that possibility."

On moving house: "I should hate to think I might have to do it some day – what with my many thousands of books, to say nothing of the stock of Arkhsm House—yet this latter will have to be moved this summer when my new warehouse is up, out of the basement and one gable room of this spacious house, and into the new quarters for a more efficient operation. That is a prospect I view with horror, esp. since I must . . . prepare two books for the printer, teach two weeks, and then go into hospital early in August [1969]–my gall bladder must be removed, they tell me."

On his family: "Yes, thank you, the children are well. April Rose will be fifteen August 9, and Walden thirteen August 22. April already helps with Arkham House, entering books so that I can pay royalties when they fall due, without delay . . . My major work,–i.e., that I want most to do, must now wait upon my return from hospital."

On his illness: [November 3rd, 1969]: "Yes, my silence has been due to illness. I am just back from hospital – eighty-seven days on my back, four operations, pneumonia, peritonitis, pleuritis, a collapsed and punctured lung, hepatitis–well, you name it. Had I been a heavy smoker or drinker, I'd have been planted by now; being neither, and keeping myself in good condition saved me. One dr. described me to another as 'a tough old bastard' which I took as a compliment. For a month my condition was critical, but now I'm back at the old stand–v. weak, learning to walk . . . I face 500 letters to answer."

One could go on quoting. Several things stand out from all this. Great physical courage; lion-heartedness in adversity; good humour; optimism; generosity and probity in business dealings. Any one of these things would outweigh the debit side of most man.

Before I forget, I should add a few more things to the list of his achievements: devoted father and family man; the writing of children's books; the filming of his works for TV and the cinema; a new venture, the recording of his own poems, read by himself; his prolific lecturing stint at universities in Wisconsin, where he conducted seminars; walking; swimming; the writing of detective stories, which included the Judge Peck series; chess; and the collecting of a library of over 12,000 volumes, with special emphasis on fantasy and the macabre, of course.

Among his many honours were the award of the Guggenheim Fellowship in 1938 and recognition from a large number of universities and organisations followed. He received the Apostolic Blessing of Pope John for his Wisconsin Books series in 1959.

In my mind's eye I still see him, a kindly and gigantic figure, striding along beside the shore of Walden Pond or giving up the whole of May every year, wandering the woods in search of his beloved morels. And his cheerful salutation, at the end of his letters, in which one seemed almost to hear the voice: "All best, always. Cordially yours."

These are clichés, I know, but true just the same. We shall not see his like again. He will be missed through the years, not only by me to whom he was a friend and an encourager, but by countless thousands whom he had helped by his generosity, his example, his courage and his talent. All best, always, August.

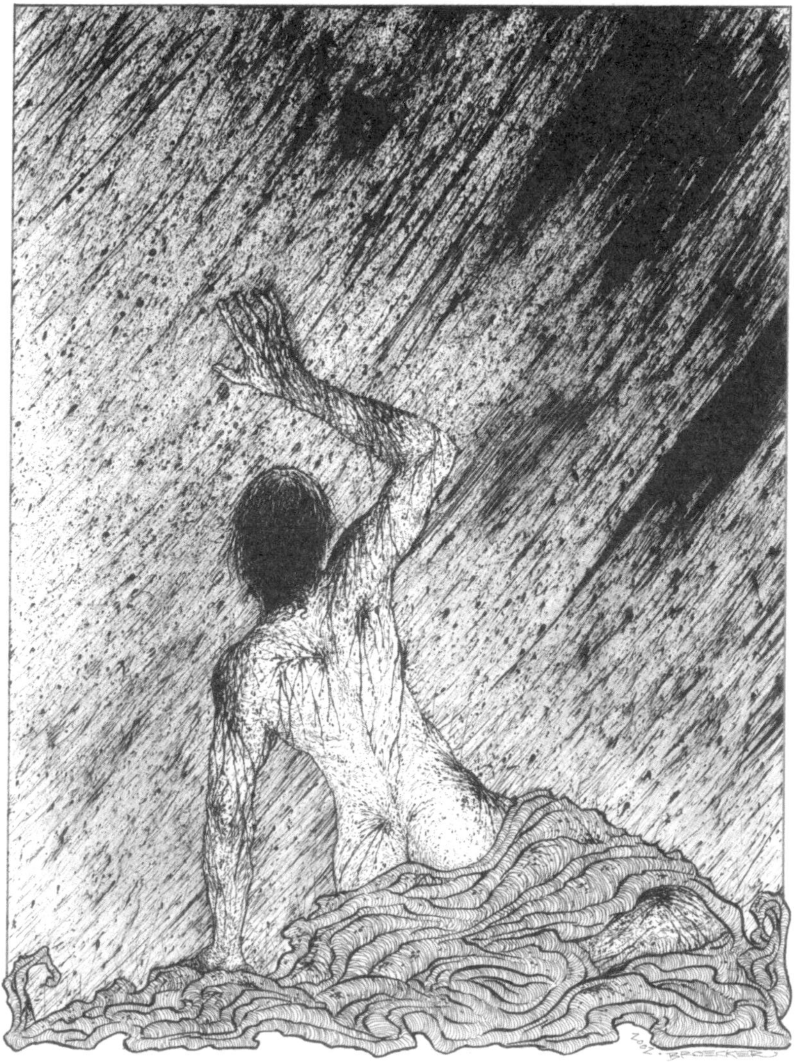

LES EDWARDS.

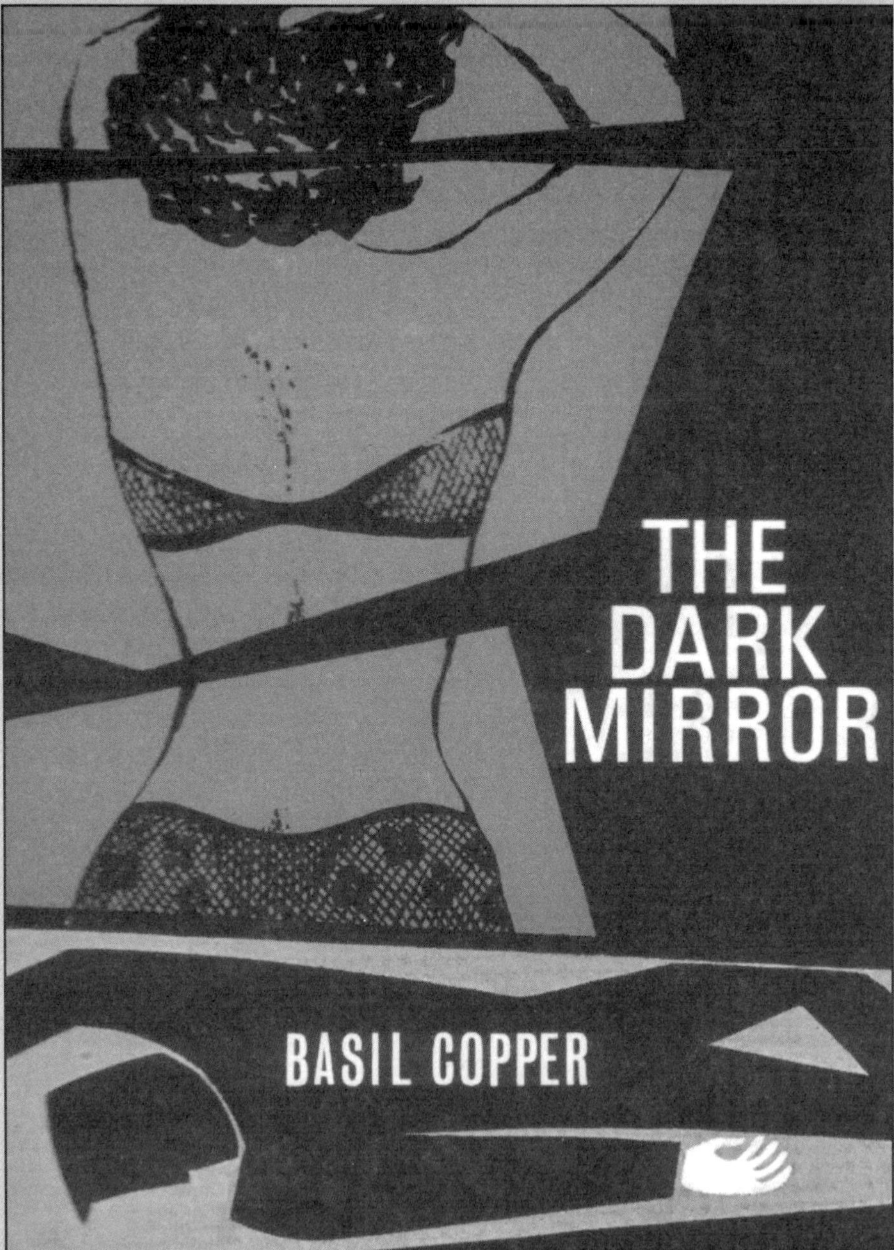
THE
DARK
MIRROR
BASIL COPPER

D. THE "MIKE FARADAY" SERIES

Includes an alphabetical listing of novels written by Basil Copper

(*"The pure detective story can emerge from any innocuous opening that the writer cares to imagine and from there the plot can slowly crystallise as the author goes on. Here I speak purely for myself because I know that many authors, both famous and obscure, meticulously plot their detective novels so that they know what is going to happen right from page one.*

"That is not my method; mine is the same as that of one of the great masters of the American thick-ear novel, Raymond Chandler. Chandler's dictum was to say that when one got stuck, have a man with a gun come through the door. This always works in my case and sends the plot off at another tangent." BC.)

D1 **BAD SCENE** (#50)

a. __________, Robert Hale Limited, London, England, October 1987 (hc)
£10.25 / 192pp / Jacket design by Colin Andrews.
b. [as **FARLIGA VATTEN**], WinPocket/Winther, Trondheim, Sweden 1988 (pb).
c. __________, F.A. Thorpe (Publishing)/Linford Mystery Library
[Large Print Edition], Leicester, England, February 1991 (tp).
d. __________, Chivers Large Print/G.K. Hall & Co., Bath, England, 2000 (hc).
e. __________, Camden Large Print/G.K. Hall & Co., Bath, England, 2000 (tp).
f. __________, Nightingale (Large Print)/G.K. Hall & Co., Maine, USA, 2000 (tp).

D2 **THE BIG CHILL** (#10)

a. __________, Robert Hale & Company, London, England, 1972 (hc)
Price: £1.50 / 183pp / Jacket design by Colin Andrews.
b. [as **DER SCHWARZE RITTER**], Wilhelm Goldmann Verlag/Krimi Verlag AG
Wollerau, West Germany/Switzerland, [c1972] (pb).
c. [as **KALD SOM STÅL**], Forlaget For Alle, Trondheim, Norway, 1983 (pb).
[Note: "Mike Farady #12"]
d. [as **CHLADNY DECH SMRTI**], see **TRI PRÍPADY A MIKE FARADAY** [D48].
e. __________, F.A. Thorpe (Publishing)/Linford Mystery Library
[Large Print Edition], Leicester, England, 2001 (tp).

D3 **THE BIG RIP-OFF** (#28)

a. _________, Robert Hale Limited, London, England, July 1979 (hc)
Price: £4.15 / 158pp / Jacket design by Colin Andrews.
b. [as **GELD SPIELT (K)EIN ROLLE**], Wilhelm Goldmann Varlag, Munich, West Germany, September 1980 (pb).
c. [as **TRIPLICE OLOCAUSTO PER MIKE**], Giallo D'Azione Mondadori, Milan, Italy, February 1981 (pb).
d. [as **HEVNEN ER SØT**], Forlaget For Alle, Trondheim, Norway, 1984 (pb).
[Note: "Mike Faraday #15"]
e. _________, F.A. Thorpe (Publishing) Limited/Linford Mystery Library [Large Print Edition], Leicester, England, 2002 (tp).

D4 **BLOOD ON THE MOON** (#47)

a. _________, Robert Hale Limited/Connoisseur Crime, London, England, 1986 (hc)
£7.95 / 172pp.
b. _________, F.A. Thorpe (Publishing) Limited/Linford Mystery Library [Large Print Edition], Leicester, England, 2004 (tp).

D5 **THE BREAKING POINT** (#14)

a. _________, Robert Hale & Company, London, England, 1973 (hc)
Price: £1.60 / 188pp / Jacket design by Colin Andrews.
b. [as **MIKE FARADAYS VISITENKARTE**], Wilhelm Goldmann Verlag/Krimi Verlag AG Wollerau, West Germany/Switzerland, [c1974] (hc).
c. [as **LA SOFFIATA**], Longanesi & C., Milan, Italy, 1974 (pb).
d. [as **BRÄNNANDE BEVIS**], B. Wahlströms Bokförlag/Pocketdeckaren, Stockholm, Sweden, 1974 (pb).
e. _________, F.A. Thorpe (Publishing) Limited/Linford Mystery Library [Large Print Edition], Leicester, England, November 1995 (tp).

D6 **THE CALIGARI COMPLEX** (#29)

a. _________, Robert Hale Limited, London, England, February 1980 (hc)
Price: £4.75 / 160pp / Jacket design by Colin Andrews.
b. [as **FRITT FALL**], Forlaget For Alle, Trondheim, Norway, 1982 (pb).
[Note: "Mike Faraday #1"]
c. _________, F.A. Thorpe (Publishing) Limited/Linford Mystery Library [Large Print Edition], Leicester, England, 1999 (tp).

D7 **CRACK IN THE SIDEWALK** (#22)

a. __________, Robert Hale & Company, London England, June 1976 (hc) £2.95 / 183pp.

b. [as **DER GEISTERBAHN-MÖRDER**], Wilhelm Goldmann Verlag/Krimi Verlag AG Wollerau, West Germany/Switzerland, [c1976] (pb).

c. [as **DET LUKTER AV MORD**], Forlaget For Alle, Trondheim, Norway, 1983 (pb). [Note: "Mike Farady #13"]

d. __________, F.A. Thorpe (Publishing) Limited/Linford Mystery Library [Large Print Edition], Leicester, England, 1997 (tp).

D8 **DARK ENTRY** (#33)

a. __________, Robert Hale Limited, London, England, September 1981 (hc) Price: £6.25 / 158pp / Jacket photography by Colin Andrews.

b. __________, F.A. Thorpe (Publishing) Limited/Linford Mystery Library [Large Print Edition], Leicester, England, 2004 (tp).

D9 **THE DARK MIRROR** (#1)

a. __________, Robert Hale Limited, London, England, March 1966 (hc) Price: 12/6 / 192pp / Jacket design by Michael Hoare. [Note: "In this first of a series, Californian private eye Mike Faraday reckons the case is routine, until a silenced gun cuts down Horvis the antique dealer, and involves Mike in a trail of murder and violence."]

b. [as **BÄDDAT FÖR MORD**], Kometförlaget, Falun, Sweden, 1968 (pb).

c. [as **DER DUNKLE SPIEGEL**], Wilhelm Goldmann Verlag/Krimi Verlag AG Wollerau, West Germany/Switzerland, [c1972] (pb).

d. [as **DEN USYNLIGE MORDER**], Forlaget For Alle, Trondheim, Norway, 1982 (pb). [Note: "Mike Farady #4"]

e. __________, Chivers Press/Black Dagger Crime, England, 1991 (hc).

f. __________, F.A. Thorpe (Publishing) Limited/Linford Mystery Library [Large Print Edition], Leicester, England, 1997 (tp).

D10 **DEAD FILE** (#8)

a. __________, Robert Hale & Company, London, England, 1970 (hc) Price: 21/- (£1.05) / 186pp / Jacket design by Colin Andrews.

b. [as **PÅ EGEN RISK**], Kometförlaget, Falun, Sweden, 1971 (pb).

c. [as **L'INAFFERRABILE MAURITIUS**], Longanesi & C., Milan, Italy, 1971 (pb).

d. [as **DUELL ZU DRITT**], Wilhelm Goldmann Verlag/Krimi Verlag AG Wollerau, West Germany/Switzerland, [c1972] (pb).

e. __________, F.A. Thorpe (Publishing)/Linford Mystery Library [Large Print Edition], Leicester, England, January 1991 (tp).

D11 **DEATH SQUAD** (#25)

a. __________, Robert Hale Limited, London, England, May 1977 (hc)

Price: £3.15 / 172pp / Jacket design by Colin Andrews.

b. [as **EIN TOTER MUß STERBEN**], Wilhelm Goldmann Verlag, West Germany, June 1978 (pb).

c. [as **DÖDLIGT DUBBEL SPEL**], Kometdeckaren, Falun, Sweden, 1978 (pb).

d. [as **DØDSPATRULJEN**], Forlaget For Alle, Trondheim, Norway, 1983 (pb).
[Note: "Mike Farady #11"]

e. __________, F.A. Thorpe (Publishing) Limited/Linford Mystery Library [Large Print Edition], Leicester, England, 1999 (tp).

D12 **DIE NOW, LIVE LATER** (#5)

a. __________, Robert Hale Limited., London, England, January 1968 (hc)

Price: 16/- / 192pp / Jacket design by Colin Andrews.

b. [as **MIKE OG DYBFRYSERLIGENE**], Ark's Forlag, Copenhagen, Denmark, 1968 (pb).

c. [as **ISKALL BLUFF**], B. Wahlströms Bokförlag/Pocketdeckaren, Stockholm, Sweden, 1973 (pb).

d. __________, F.A. Thorpe (Publishing) Limited/Linford Mystery Library [Large Print Edition], Leicester, England, January 1993 (tp).

D13 **DON'T BLEED ON ME** (#6)

a. __________, Robert Hale, London, England, 1968 (hc)

Price: 16/- / 192pp / Jacket design by Colin Andrews.

b. [as **LAD IKKE BLODET DRYPPE PÅ MIG**], Ark's Forlag, Copenhagen, Denmark, 1968 (pb).

c. [as **SKYLLA SIG SJÄLV**], B. Wahlströms Bokförlag/Pocketdeckaren, Stockholm, Sweden, 1973 (pb).

d. [as **DØDE MENN SNAKKER IKKE**], Forlaget For Alle, Trondheim, Norway, 1983 (pb).
[Note: "Mike Farady #8"]

e. __________, F.A. Thorpe (Publishing)/Linford Mystery Library [Large Print Edition], Leicester, England, September 1991 (tp).

f. __________, Chivers Large Print/Thorndike Press, Bath,

England, 2005 (hc).
g. __________, Camden Large Print/Thorndike Press, Bath, England, 2005 (tp).
h. __________, British Favourites (Large Print)/Thorndike Press, Maine, USA, 2005 (tp).

D14 **THE EMPTY SILENCE** (#32)
a. __________, Robert Hale Limited, London, England, 1981 (hc)
Price: £5.95 / 159pp / Jacket design by Colin Andrews.
b. __________, F.A. Thorpe (Publishing) Limited/Linford Mystery Library [Large Print Edition], Leicester, England, August 1994 (tp).

D15 **THE FAR HORIZON** (#36)
a. __________, Robert Hale Limited, London, England, 1982 (hc)
Price: £6.75 / 188pp / Jacket design by Colin Andrews.
b. __________, F.A. Thorpe (Publishing) Limited/Linford Mystery Library [Large Print Edition], Leicester, England, May 1993 (tp).
c. __________, Chivers Large Print/G.K. Hall & Co., Bath, England, 2001 (hc).
d. __________, Camden Large Print/G.K. Hall & Co., Bath, England, 2001 (tp).
e. __________, Nightingale (Large Print)/G.K. Hall & Co., Maine, USA, 2001 (tp).

D16 **THE FAR SIDE OF FEAR** (#44)
a. __________, Robert Hale Limited, London, England, September 1985 (hc)
Price: £7.75 / 190pp / Jacket design by Colin Andrews.
b. __________, F.A. Thorpe (Publishing)/Linford Mystery Library [Large Print Edition], Leicester, England, 2005 (tp).

D17 **FEEDBACK** (#16)
a. __________, Robert Hale & Company, London, England, June 1974 (hc)
Price: £1.75 / 184pp / [Jacket design by Colin Andrews].
b. [as **MORDVERSUCH IN STUDIO B**], Wilhelm Goldmann Verlag/Krimi Verlag AG Wollerau, West Germany/Switzerland, [c1974] (hc).
c. __________, F.A. Thorpe (Publishing)/Linford Mystery Library [Large Print Edition], Leicester, England, November 1991 (tp).
d. __________, Chivers Large Print/G.K. Hall & Co., Bath, England, 2001 (hc).
e. __________, Camden Large Print/G.K. Hall & Co., Bath, England, 2001 (tp).
f. __________, Nightingale (Large Print)/G.K. Hall & Co., Waterville, Maine, USA, 2001 (tp).

D18 **FLIP-SIDE** (#30)

a. _________, Robert Hale Limited, London, England, November 1980 (hc).
Price: £5.50 / 172pp / Jacket photography by Colin Andrews.
b. [as **BLINDSPOR**], Forlaget For Alle, Trondheim, Norway, 1982 (pb).
[Note: "Mike Farady #6"]
c. [as **SCHWARZES GIFT – BLOND SERVIERT**], Wilhelm Goldman Verlag, Munich, West Germany, December 1982 (pb).
d. _________, F.A. Thorpe (Publishing)/Linford Mystery Library [Large Print Edition], Leicester, England, 2002 (tp).

D19 **A GOOD PLACE TO DIE** (#20)

a. _________, Robert Hale & Company, London, England, 1975 (hc)
Price: £2.40 / 182pp.
b. [as **EIN BOURBON FÜR DIE NERVEN**], Wilhelm Goldmann Verlag/Krimi Verlag AG Wollerau, West Germany/Switzerland, [c1975] (hc).
c. [as **EIN BOURBON FÜR DIE NERVEN**], Wilhelm Goldmann Verlag/Krimi Verlag AG Wollerau, West Germany/Switzerland, [c1975] (pb).
d. _________, F.A. Thorpe (Publishing)/Linford Mystery Library [Large Print Edition], Leicester, England, September 1989 (tp).

D20 **A GREAT YEAR FOR DYING** (#12)

a. _________, Robert Hale & Company, London, England, March 1973 (hc)
Price: £1.50 / 190pp / Jacket design by Colin Andrews.
b. [as **EIN GUTES JAHR ZUM STERBEN**], Wilhelm Goldmann Verlag/Krimi Verlag AG Wollerau, West Germany/Switzerland, [c1973] (hc).
c. [as **EIN GUTES JAHR ZUM STERBEN**], Wilhelm Goldmann Verlag/Krimi Verlag AG Wollerau, West Germany/Switzerland, [c1973] (pb).
d. [as **MÖRDANDE HOT**], B. Wahlströms Bokförlag/Pocketdeckaren, Stockholm, Sweden, 1974 (pb).
e. _________, F.A. Thorpe (Publishing)/Linford Mystery Library [Large Print Edition], Leicester, England, March 1993 (tp).

D21 **HANG LOOSE** (#34)

a. _________, Robert Hale Limited, London, England, February 1982 (hc)
Price: £6.50 / 160pp / Jacket design by Colin Andrews.
b. _________, Chivers Press/A Lythway Book [Large Print Edition], Bath, England, 1990 (hc).
c. _________, F.A. Thorpe (Publishing)/Linford Mystery Library

[Large Print Edition], Leicester, England, 2004 (tp).

D22 **HARD CONTRACT** (#40)

a. _________, Robert Hale Limited, London, England, January 1984 (hc)
Price: £7.50 / 157pp / Jacket design by Colin Andrews.
b. _________, F.A. Thorpe (Publishing)/Linford Mystery Library
[Large Print Edition], Leicester, England, February 1988 (tp).

D23 **HEAVY IRON** (#48)

a. _________, Robert Hale Limited/Connoisseur Crime, London, England, 1987 (hc)
Price: £7.95 / 159pp.
b. _________, F.A. Thorpe (Publishing) Limited/Linford Mystery Library
[Large Print Edition], Leicester, England, 1999 (tp).
c. _________, Chivers Large Print/Thorndike Press, Bath, England, 2003 (hc).
d. _________, Camden Large Print/Thorndike Press, Bath, England, 2003 (tp).
e. _________, Nightingale (Large Print) /Thorndike Press, Maine, USA, 2003 (tp).

D24 **THE HIGH WALL** (#18)

a. _________, Robert Hale & Company, London, England, February 1975 (hc)
Price: £1.95 / 182pp / Jacket design by Colin Andrews.
b. [as **DIE MAUER DES SCHWEIGENS**], Wilhelm Goldmann Verlag/Krimi Verlag AG Wollerau, West Germany/Switzerland, [c1973] (pb).

c. [as **MASKERAT BROTT**], Kometdeckaren, Falun, Sweden, 1976 (pb).
d. _________, F.A. Thorpe (Publishing) Limited/Linford Mystery Library [Large Print Edition], Leicester, England, 1987 (tp).
e. _________, Chivers Large Print/Thorndike Press, Bath, England, 2006 (hc).
f. _________, Camden Large Print/Thorndike Press, Bath, England, 2006 (tp).
g. _________, British Favourites (Large Print)/Thorndike Press, Maine, USA, 2006 (tp).

D25 **THE HOOK** (#41)

a. _________, Robert Hale Limited, London, England, 1984 (hc)
Price: £7.50 / 192pp / Jacket design by Colin Andrews.
b. _________, F.A. Thorpe (Publishing) Limited/Linford Mystery Library
[Large Print Edition], Leicester, England, 1995 (tp).

D26 **HOUSE-DICK** (#51)

a. __________, Robert Hale Limited, London, England, 1988 (hc)
Price: £10.50 / 158pp / Jacket design by Colin Andrews.

b. __________, F.A. Thorpe (Publishing)/Linford Mystery Library [Large Print Edition], Leicester, England, 2000 (tp).

D27 **IMPACT** (#19)

a. __________, Robert Hale & Company, London, England, June 1975 (hc)
£2.25 / 187pp / Jacket design by Colin Andrews.

b. [as **DIE BOMBENSICHERE ERBSCHAFT**], Wilhelm Goldmann Verlag/Krimi Verlag AG Wollerau, West Germany/Switzerland, [c1975] (hc).

c. [as **DIE BOMBENSICHERE ERBSCHAFT**], Wilhelm Goldmann Verlag/Krimi Verlag AG Wollerau, West Germany/Switzerland, [c1975] (pb).

d. __________, F.A. Thorpe (Publishing) Limited/Linford Mystery Library [Large Print Edition], Leicester, England, 1997 (tp).

D28 **JET-LAG** (#46)

a. __________, Robert Hale Limited, London, England, 1986 (hc)
Price: £8.50 / 192pp / Jacket Photograph by Colin Andrews.

b. __________, F.A. Thorpe (Publishing)/Linford Mystery Library [Large Print Edition], Leicester, England, January 1990 (tp).

c. __________, Chivers Large Print/Thorndike Press, Bath, England, 2004 (hc).

d. __________, Camden Large Print/Thorndike Press, Bath, England, 2004 (tp).

e. __________, Nightingale (Large Print) /Thorndike Press, Maine, USA, 2004 (tp).

D29 **THE LONELY PLACE** (#21)

a. __________, Robert Hale & Company, London, England, February 1976 (hc)
Price: £2.80 / 184pp.

b. [as **IN SCHLECHTER GESELLSCHAFT**], Wilhelm Goldmann Verlag/Krimi Verlag AG Wollerau, West Germany/Switzerland, [c1976] (hc).

c. __________, F.A. Thorpe (Publishing) Limited/Linford Mystery Library [Large Print Edition], Leicester, England, 1997 (tp).

D30 **THE LONG REST** (#31)

a. __________, Robert Hale Limited, London, England, January 1981 (hc)
Price: £5.75 / 160pp / Jacket design by Colin Andrews.

b. [as **BAKHOLD**], Forlaget For Alle, Trondheim, Norway, 1982 (pb).
[Note: "Mike Farady #3"]

c. __________, Chivers Large Print/G.K. Hall & Co., Bath, England, 1994 (hc).
d. __________, Camden Large Print/G.K. Hall & Co., Bath, England, 1994 (tp).
e. __________, Nightingale (Large Print)/G.K. Hall & Co., Thorndike, Maine, USA, 1994 (tp).
f. __________, F.A. Thorpe (Publishing) Limited/Linford Mystery Library [Large Print Edition], Leicester, England, 2003 (tp).

D31 **THE MARBLE ORCHARD** (#7)

a. __________, Robert Hale, London, England, September 1969 (hc)
Price: 18/- (90p) / 159pp / Jacket design by Colin Andrews.
b. [as **VEJEN TIL GRAVEN**], Arks Forlag, Copenhagen, Denmark, [no date] (pb).
c. [as **FAST I NÄTET**], Kometförlaget, Falun, Sweden, 1971 (pb).
d. [as **SECHS PATRONEN IM MAGAZIN**], Wilhelm Goldman Verlag/Krimi Verlag AG Wollerau, West Germany/Switzerland, 1972 (pb).

e. __________, F.A. Thorpe (Publishing) Limited/Linford Mystery Library [Large Print Edition], Leicester, England, 1998 (tp).

D32 **MURDER ONE** (#26)

a. __________, Robert Hale Limited, London, England, February 1978 (hc)
Price: £3.50 / 176pp / Jacket design by Colin Andrews.
b. [as **MORD ERSTEN GRADES**], Wilhelm Goldman Verlag, Munich, West Germany, July 1980 (pb).
c. [as **GRAV ÅT ANDRA**], Kometdeckaren, Falun, Sweden, 1981 (pb).
d. [as **DEN SOM GRAVER EN GRAV . . .**], Forlaget For Alle, Trondheim, Norway, 1983 (pb).
[Note: "Mike Faraday #10"]
e. __________, F.A. Thorpe (Publishing)/Linford Mystery Library [Large Print Edition], Leicester, England, 2002 (tp).

D33 **THE NARROW CORNER** (#39)

a. __________, Robert Hale Limited, London, England, September 1983 (hc)
Price: £7.25 / 187pp / Jacket design by Colin Andrews.

b. __________, F.A. Thorpe (Publishing) Limited/Linford Mystery Library [Large Print Edition], Leicester, England, November 1989 (tp).

D34 **NIGHT FROST** (#2)

a. __________, Robert Hale Limited, London, England, 1966 (hc)
Price: 13/6 / 192pp / Jacket design by Colin Andrews.

b. [as **NACHTFROST**], Die roten Goldman Krimi, West Germany, 1966 (pb).

c. [as **MIKE PÅ BAHAMA**], Ark's Forlag, Copenhagen, Denmark, 1968 (pb).

d. [as **GHIACCIO AL TROPICO FUOCO NELL'ACQUA**], Longanesi & C., Milan, Italy, 1971 (pb).

e. [as **PIRAJA**], Forlaget For Alle, Trondheim, Norway, 1982 (pb).
[Note: "Mike Faraday #2"]

f. __________, F.A. Thorpe (Publishing) Limited/Linford Mystery Library [Large Print Edition], Leicester, England, 1996 (tp).

g. __________, Chivers Press/Black Dagger Crime, Bath, England, 2000 (hc).

D35 **NO FLOWERS FOR THE GENERAL** (#3)

a. __________, Robert Hale Limited, London, England, January 1967 (hc)
Price: 15/- / 190pp / Jacket design by Colin Andrews.

b. [as **MED ORDER ATT DÖDA**], Kometförlaget, Falun, Sweden, 1968 (pb).

c. [as **MIKE NÅR ALT**], Ark's Forlag, Copenhagen, Denmark, 1968 (pb).

d. [as **KEINE BLUMEN FÜR DEN GENERAL**], Wilhelm Goldmann Verlag/Krimi Verlag AG Wollerau, West Germany/Switzerland, [c1972] (pb).
[Note: The cover of this book features a black and white photo of Vincent Price from *The Oblong Box*.]

e. [as **GENERALEN ØNSKER IKKE BLOMSTER**], Forlaget For Alle, Trondheim, Norway, 1982 (pb).
[Note: "Mike Farady #5"]

f. __________, F.A. Thorpe (Publishing)/Linford Mystery Library [Large Print Edirion], Leicester, England, 2000 (tp).

g. __________, Chivers Press/Black Dagger Crime, Bath, England, 2001 (hc).

D36 **NO LETTERS FROM THE GRAVE** (#9)

a. __________, Robert Hale & Company, London, England, November 1971 (hc)
Price: £1.20 / 160pp / Jacket design by Colin Andrews.
b. [as **INGA BREV FRÅN GRAVEN**], B. Wahlströms Bokförlag/Pocketdeckaren, Stockholm, Sweden, 1972 (pb).
c. [as **ERSTE TÜR RECHTS**], Wilhelm Goldmann Verlag/Krimi Verlag AG Wollerau, West Germany/Switzerland, [c1972] (pb).
d. [as **ZÁDNÉ ZPRÁVY ZE ZÁHROBI**], see **TRI PRÍPADY A MIKE FARADAY** [D48].
e. __________, F.A. Thorpe (Publishing) Limited/Linford Mystery Library [Large Print Edition], Leicester, England, 1998 (tp).

D37 **PRESSURE-POINT** (#38)

a. __________, Robert Hale Limited, London, England, 1983 (hc)
£7.25 / 188pp / Jacket design by Colin Andrews.
b. __________, F.A. Thorpe (Publishing) Limited/Linford Mystery Library [Large Print Edition], Leicester, England, October 1991 (tp).

D38 **PRINT-OUT** (#52)

a. __________, Robert Hale Limited, London, England, 1988 (hc)
Price: £9.95 / 175pp / Jacket design by Colin Andrews.
b. __________, Dales Large Print Books, North Yorkshire, England, 1993 (tp).
c. __________, F.A. Thorpe (Publishing) Limited/Linford Mystery Library [Large Print Edition], Leicester, England, 1996 (tp).

D39 **A QUIET ROOM IN HELL** (#27)

a. __________, Robert Hale Limited, London, England, March 1979 (hc)
Price: £3.95 / 159pp / Jacket design by Malcolm W. Grealy.
b. [as **SKUGGAD AV DÖDEN**], Kometdeckaren, Falun, Sweden, 1982 (pb).
c. [as **DET HVITE HELVETE**], Forlaget For Alle, Trondheim, Norway, 1983 (pb).
[Note: "Mike Farady #9"]
d. __________, F.A. Thorpe (Publishing) Limited/Linford Mystery Library [Large Print Edition], Leicester, England, 1998 (tp).

D40 **RICOCHET** (#17)

a. __________, Robert Hale, London, England, 1974 (hc)
Price: £1.70 / 158pp / Jacket design by Colin Andrews.
b. [as **UNSCHULD IM REGEN**], Wilhelm Goldmann Verlag/Krimi Verlag AG Wollerau, West Germany/Switzerland, [c1973] (hc).

c. [as **UNSCHULD IM REGEN**], Wilhelm Goldmann Verlag/Krimi Verlag AG Wollerau, West Germany/Switzerland, [c1973] (pb).
d. [as **MÅNGA OM BYTET**], Kometdeckaren, Falun, Sweden, 1976 (pb).
e. [as **RIKOSJETT**], Forlaget For Alle, Trondheim, Norway, 1982 (pb).
[Note: "Mike Farady #7"]
f. _________, F.A. Thorpe (Publishing)/Linford Mystery Library [Large Print Edition], Leicester, England, June 1993 (tp).
g. [as **BUMERANG**], see **TRI PRÍPADY A MIKE FARADAY** [D48].

D41 **SCRATCH ON THE DARK** (#4)

a. _________, Robert Hale Limited, London, England, September 1967 (hc)
Price: 15/- / 190pp / Jacket design by Colin Andrews.
b. [as **MIKE SKAL MYRDES**], Ark's Forlag, Copenhagen, Denmark, 1968 (pb).
c. [as **NAKEN SKRÄCK**], Kometförlaget, Falun, Sweden, 1969 (pb).
d. [as **UN GRAFFIO NEL BUIO**], Longanesi & C., Milan, Italy, 1970 (pb).
e. _________, F.A. Thorpe (Publishing)/Linford Mystery Library [Large Print Edition], Leicester, England, 2000 (tp).
f. _________, Chivers Press/Black Dagger Crime, Bath, England, 2002 (hc).

D42 **SHOCK-WAVE** (#13)
a. _________, Robert Hale & Company, London, England, July 1973 (hc)
Price: £1.50 / 184pp / Jacket design by Colin Andrews.
b. [as **DER GEHEIMNISVOLLE MR. CHOU**], Wilhelm Goldmann Verlag/Krimi Verlag AG Wollerau, West Germany/Switzerland, [c1973] (hc).

c. _________, IPC Magazines/Fleetway Thriller Library #2, London, England, 1977 (magazine).
d. _________, F.A. Thorpe (Publishing) Limited/Linford Mystery Library [Large Print Edition], Leicester, England, October 1994 (tp).

D43 **SHOOT-OUT** (#35)
a. _________, Robert Hale Limited, London, England, May 1982 (hc)
Price: £6.75 / 173pp / Jacket photography by Colin Andrews.
b. _________, F.A. Thorpe (Publishing) Limited/Linford Mystery Library

[Large Print Edition], Leicester, England, 2005 (tp).

D44 **SNOW-JOB** (#45)

a. __________, Robert Hale Limited, London, England, 1986 (hc)
Price: £7.95 / 187pp / Jacket photography by Colin Andrews.
b. __________, F.A. Thorpe (Publishing)/Linford Mystery Library [Large Print Edition], Leicester, England, March 1989 (tp).
c. [as **ÄLÄ HUOLI HUOMISESTA**], Book Studio, Hyvinkää, Finland, November 1992 (pb).

D45 **STRONG-ARM** (#11)

a. __________, Robert Hale, London, England, 1972 (hc)
Price: £1.50 / 183pp / Jacket design by Colin Andrews.
b. [as **TOD IN DER GALERIE**], Wilhelm Goldmann Verlag/Krimi Verlag AG Wollerau, West Germany/Switzerland, 1972 (hc).
c. [as **TOD IN DER GALERIE**], Wilhelm Goldmann Verlag/Krimi Verlag AG Wollerau, West Germany/Switzerland, 1972 (pb).
d. [as **UN BRACCIO DI FERRO**], Longanesi & C., Milan, Italy, 1973 (pb).
e. __________, F.A. Thorpe (Publishing)/Linford Mystery Library [Large Print Edition], Leicester, England, January 1989 (tp).

D46 **TIGHT CORNER** (#23)

a. __________, Robert Hale & Company, London, England, October 1976 (hc)
£2.95 / 157pp.
b. __________, F.A. Thorpe (Publishing)/Linford Mystery Library [Large Print Edition], Leicester, England, July 1994 (tp).

D47 **TRIGGER-MAN** (#37)

a. __________, Robert Hale Limited, London, England, January 1983 (hc)
Price: £6.95 / 189pp / Jacket design by Colin Andrews.
b. __________, F.A. Thorpe (Publishing)/Linford Mystery Library [Large Print Edition], Leicester, England, June 1994 (tp).

D48 **TRI PRÍPADY A MIKE FARADAY**

a. __________, Oddych, Strelnicni, Czechoslovakia, 1995 (hc).
[Note: Omnibus edition containing the novels *Ricochet* (D40), *The Big Chill* (D10) and *No Letters from the Grave* (D36).]

D49 **TURN DOWN AN EMPTY GLASS** (#49)

a. __________, Robert Hale Limited/Connoisseur Crime, London, England, 1987 (hc) £7.95 / 176pp.

b. __________, F.A. Thorpe (Publishing) Limited/Linford Mystery Library [Large Print Edition], Leicester, England, 1999 (tp).

D50 **TUXEDO PARK** (#43)

a. __________, Robert Hale Limited, England, May 1985 (hc) Price: £7.75 / 188pp / Jacket photography by Colin Andrews.

b. __________, F.A. Thorpe (Publishing)/Linford Mystery Library [Large Print Edition], Leicester, England, March 1990 (tp).

c. __________, Chivers Large Print/G.K. Hall & Co., Bath, England, 2004 (hc).

d. __________, Camden Large Print/G.K. Hall & Co., Bath, England, 2004 (tp).

e. __________, Nightingale (Large Print)/G.K. Hall & Co., Maine, USA, 2004 (tp).

D51 **A VOICE FROM THE DEAD** (#15)

a. __________, Robert Hale & Company, London England, 1974 (hc) Price: £1.60 / 187pp / Jacket design by Colin Andrews.

b. [as **DER ACHT-STUNDEN-FALL**], Wilhelm Goldmann Verlag/Krimi Verlag AG Wollerau, West Germany/Switzerland, [c1975] (hc).

c. [as **SOM DRACULAS BORG**], Forlaget For Alle, Trondheim, Norway, 1984 (pb). [Note: "Mike Farady #16"]

d. __________, F.A. Thorpe (Publishing) Limited/Linford Mystery Library [Large Print Edition], Leicester, England, 1998 (tp).

D52 **THE YEAR OF THE DRAGON** (#24)

a. __________, Robert Hale Limited, London, England, 1977 (hc) £3.15 / 157pp / Jacket photography by Colin Andrews.

b. [as **DER EIDECHSENMANN**], Wilhelm Goldmann Verlag, West Germany, 1978 (pb).

c. [as **DRAKENS ÅR**], Forlaget For Alle, Trondheim, Norway, 1983 (pb). [Note: "Mike Farady #14"]

d. __________, F.A. Thorpe (Publishing)/Linford Mystery Library [Large Print Edition], Leicester, England, August 1991 (tp).

D53 **YOU ONLY DIE ONCE** (#42)

a. __________, Robert Hale Limited, London, England, September 1984 (hc)
Price: £7.50 / 192pp / Jacket design by Colin Andrews.

b. __________, F.A. Thorpe (Publishing)/Linford Mystery Library
[Large Print Edition], Leicester, England, 2000 (tp).

c. __________, Chivers Large Print/Thorndike Press, Bath, England, 2006 (hc).

d. __________, Camden Large Print/Thorndike Press, Bath, England, 2006 (tp).

e. __________, British Favourites (Large Print)]/Thorndike Press, Maine, USA, 2006 (tp).

(*"I am occasionally asked if I know the last page of a thriller when I begin. Let me confess here and now that I have no more idea what is to happen when I sit down to write a 60,000 word Mike Faraday novel.*

"Mike is usually sitting in his office, flat broke and waiting for something to happen. Eventually the door opens, a client enters, ushered in by his blonde secretary Stella, and the story begins.

"Raymond Chandler, who has influenced me greatly, once sagely observed, 'When you're stuck for ideas have a man with a gun come through the door'. This always works beautifully, and along around the fourth chapter the plot is beginning to emerge.

"As I go along I type plot points and further ideas on separate sheets of paper and eventually everything is crystal clear–except to the author! The most tricky thing in writing such a thriller is finishing up. When you realise you are getting close to the end and you have to kill off the villain and explain the whole plot in ten pages, it's quite an effort, believe me. Especially when you loathe the idea of everyone sitting around a table while the private eye explains the whole set-up. I always avoid that and try to convey the situation through terse conversation." BC.)

RAP-SHEET

[NOTE: The last Mike Faraday novel was *Print-Out*, published by Robert Hale Limited in 1988. On May 14, 1995, Basil Copper began work on the opening of a new Faraday novel ("Suspicious people come to view coffin. Mobsters."), which would have been #53 in the series. The following page is dated 18/11/2001.]

by Basil Copper

RAP-SHEET

IT WAS 10:50 a.m. when I hit the crematorium. Abel Kipling was getting a great send-off. There were all of three people there: the preacher; the mortician; and the guy who operated the furnace.

I'd been hired by a woman called Kiki von Brinken to keep an eye on things. Just why, I wasn't quite sure. The casket was on a trolley ready to go up a ramp at the side entrance when I showed up. At least he had a nice sunny day for it, I thought.

Just then a little guy with hard eyes and a rat-trap mouth materialised at my elbow. He was wearing a loud check suit so I guessed he wasn't actually a mourner. He was shifting gear on some gum in his mouth so I didn't interrupt his lunch. He glanced at the bleak surroundings, watching the preparations for wheeling the trolley up.

"The arseholes of the world," he said softly.

"You could say that," I told him.

He shifted some more gum, looking out across the expanse of the Garden of Remembrance where a nice-looking girl was studying the metal plaques intently.

"That blonde gives me balls a headache," he said.

At that point I figured he wasn't a local Baptist minister in mufti, either.

"Oughtn't you be helping them up with the trolley," I said.

He gave me a sharp look. "Who the hell do you think I am?"

"Certainly not the editor of *Good Taste* magazine," I said.

He gave me a twisted smile. "You can say that again."

I obliged him and a scowl spread across his unattractive features. He shifted from one foot to the other, still keeping an eye on the receding tableau, the clergyman intoning a muffled prayer.

"Private dick?" he hinted.

"Police department?" I said.

The frozen look was back again. "I ask the questions."

I called his bluff. "It's a free country," I said. "Let's see your badge."

He smiled then. Genuinely this time. "You've got a nerve, mister. I'll say that for you."

I didn't know what the hell he was talking about, but I let him ramble on. He got out a comb then and carefully re-arranged the dandruff in his hair.

In 1981, Basil Copper was elected Chairman of the Crime Writers' Association of Great Britain. Seen here addressing members and guests at the annual dinner at a London hotel. He served on the committee for a total of seven years, and has been a member of the Association for forty years.

DIAMONDS ARE FOR NEVER

I

IT WAS AROUND 10:00 a.m. when I tooled my five-year-old powder-blue Buick into a parking lot in the forecourt of the DeSoto Building. I got out, the sun hot and heavy on my head and slammed the door. The heavy thud sounded like my bank balance hitting the sidewalk. It was in free fall and trying to crawl under a manhole cover so I'd be glad of the paper. I took my size nines up the front steps and across a lobby only slightly smaller than the one in the United Nations building.

A braided commissionaire stopped me before I got to the elevators.

"Sorry, sir, there's something wrong with the driving mechanisms. You'll have to walk up to the third floor, I'm afraid. You'll be all right after that."

"I'll have to get my truss re-blocked," I told him.

He gave me a twisted smile. "That's life," he said.

I was blown by the time I got up to the second, passing several geriatrics and a corpulent man who looked like Fatty Arbuckle on his day off. I stopped to get my breath when I got to the third level. You must be getting old, Mike, I told myself. I say that at least once on every case. Then I buttoned my way up to the twenty-fourth floor.

by Basil Copper

I was calling on Myles Van Doren. He'd rung half an hour earlier and said something about a girl who'd disappeared with some property of his and had asked me to come to his business office.

He had an entry phone on a steel-faced door that looked like something out of Fort Knox. I waited until I was told to come in. The door swung back and I went on over a carpet with a pile so thick it caught me behind the knees. Van Doren, when I got up to a desk as big as the deck of an aircraft carrier, was a heavily-built man of indeterminate age, with a bald head and large ears. He wouldn't make the beauty ads that was for sure, but I wasn't here for that.

"Good of you to come, Mr. Faraday."

He had a voice like a rusty nail file scraping across tin. If you get what I mean.

"I couldn't resist it," I said. "I like to eat."

He chuckled throatily. "You won't find me ungenerous in the matter of fees."

I was warming to him by the minute.

"You like women, Mr. Faraday–?" he said irrelevantly.

"Who doesn't?" I said. "But some are pretty unreliable."

"You can say that again."

His eves had a glazed look now.

"I want a beautiful girl of eighteen with firm buttocks and thighs."

I looked at him.

"How old are you?"

"Eighty-three."

"You don't want much," I said.

He smiled, exposing a row of jagged teeth. "I have eight million dollars."

I didn't persist. He had a point.

Van Doren suddenly sat bolt upright in his padded chair and his eyes were like gimlets now, boring into mine.

"To business, Mr. Faraday," he said in a steely voice.

"That's what I'm here for," I told him.

Then he relaxed again and his eyes had resumed their normal expression. I was finding it difficult to figure him out.

"There was a girl," he said slowly.

"There always is," I told him.

He gave me a crooked smile. "She took off with some property of mine. A diamond to be precise. I won't bother you with the carats and such-like technical details, but it was worth around half a million dollars. I want it back, Mr.

Faraday."

"I'm not surprised," I said. I felt like whistling at the amount but restrained myself. There was an odd silence while Van Doren drummed with claw-like fingers on the desk. I busied myself with the nice view of the smog at the big windows on the right-hand side of the room. I wanted for him to go on, but he said nothing.

"She got a name?" I said at last.

He nodded. "Several."

"One will do. The latest."

"Melissa Goddard. I'll give you her last known address here in L.A. but you won't find her home. I checked around."

I nodded, watching as he pulled across a leather-bound booklet and scribbled something in it. He tore the sheet off and tossed it across the blotter. I looked at it. There were so many noughts mine eyes dazzled. Lay off the Marlowe stuff, I told myself. Christopher, that is, not Philip.

Van Doren gave me another of his crooked smiles. "Are you on the case, Mr. Faraday?"

"I started yesterday," I told him.

II

I got back into the Buick and gunned off down the boulevard. The sun was burning off the smog nicely and occasionally one could admire the scenery. Although Van Doren had said the girl wasn't around any more, and it was quite a reasonable assumption, I felt her address was the best place to start. A neighbour might have some information or maybe the girl had left something behind which would give me an indication of her whereabouts. In any case I had nothing else to do today, apart from paying in my client's check in case it bounced. I gave myself a tight smile in the rear mirror and the joke, feeble as it was, lasted me the rest of the journey.

The girl lived way across town and it was almost midday before I got there, She'd been the old man's secretary for nearly a year, and had lit out almost a week ago. I asked why he hadn't reported the theft of the diamond to the police, but he said he'd only just discovered his loss. I smelt something phoney about that; maybe the diamond was illicit. Or, like so many old rich guys he had so much loot it was hard to keep track of it.

The Goddard number's address was a bungalow court on a location called Hacienda Drive. It was a pretty nice lay-out, but I wasn't here for the real estate brochure. No.44 was set back behind a manicured lawn with stone steps leading up to the front door. I saw that was ajar when I got to it. I smelt trouble then, especially as there was no reply to my ring.

I wasn't carrying the Smith-Wesson today, so I eased cautiously round the door and across the hallway. I went through several rooms but they were all empty except for the Californian sunlight. The place had been turned over by an expert. It looked like World War II. I found the Goddard number eventually. She was quite a looker. Right now she wasn't saying much.

She would have found it difficult because the thin cord round her neck was buried so deep in her skin it was hidden by the folds of flesh. She had been worked over, but I guessed she hadn't talked or the place wouldn't have been in such a mess. She was nude except for a G-string and she must have been really something when alive. Van Doren certainty knew how to pick them. For looks, of course, not reliability. I lifted her hand. It was still warm so she couldn't have been dead long. I wondered why the hell she had taken the risk of coming brick to L.A. Perhaps to get the diamond which she might have hidden in the bungalow? A fatal mistake as it turned out.

Whoever had chilled her wouldn't have hung around long, but I felt uneasy just the same. I sat down heavily in a chair at the foot of the bed and stared at the ceiling. That didn't help any. Somewhere a bluebottle buzzed in the oppressive silence. There's usually one in most of my cases. That didn't help any either. A motor started up somewhere on the boulevard and an automobile went away quietly in low gear. That could mean anything or nothing. Most likely a neighbour unless the killer had nerves of steel. That was irrelevant now anyway.

I sighed and reached out for the phone. I got on to Van Doren and gave him the good news. He wasn't too pleased, but I didn't blame him for that.

"The law will be around," I said. "All you have to do is play it straight."

He gave an odd choking noise and rang off. Anyway, I'd give him an hour or so to think up his answers. Then I got the law in.

It took me quite a while to get away. I was feeling pretty hungry by the time I'd been to the bank to deposit Van Doren's check, and it was around four by my non-waterproof watch when I got myself a sandwich and a cup of coffee at Jinty's. So it was after five before I hit the office. Stella was still at her desk typing. The gold bell of her hair shimmered under the lamps as she twisted her swivel chair to look at me.

"My, my," she said. "Another country heard from."

I grinned, crossing the room and putting the bank slip in front of her. Her eyes widened as she took in the amount.

"So we're in business?"

I nodded. "And you get paid this month too, honey."

She smiled. I could have watched it all day.

"Dirty business," I went on. "I just found a girl dead. Had a job to get away from the law."

Stella raises her eyebrows. "That's pretty good going, even for you, Mike. Have a cup of coffee and tell me all about it."

It was past seven by the time Stella left but I still sat on, turning the facts over in my mind. It seemed as though Van Doren was a rather trusting old bird because he had a large collection of diamonds and he was in the habit of showing them to the Goddard number. But then he had a penchant for young girls, as I'd already observed and even he had admitted it. On the day the big diamond went missing he'd left the room for a few minutes, leaving a tray of the stones on his desk. When he came back, the girl had lit out. I wondered again why he had waited a week before calling me in. He could have come by the diamond illegally, of course, which was the obvious answer, but it was none of my business anyway. And the assignment had come at the right time.

The phone rang again while the cogs of my brain were still turning over slowly. It was Van Doren once more.

"Mr. Faraday," he said in a trembling voice. "In view of what's happened are you still on the case?"

"You bet," I told him. "Like I said, whoever the killer was, he didn't get the diamond."

"Thanks again," he said. "I won't forget this."

I was thinking of shutting up shop when there came yet another call. This was getting monotonous.

It was a man's voice at the other end of the line, agitated and trembly as well. "Mr. Faraday? I need your help, sir. You wouldn't know me but my name's Nick Romano. I was Melissa Goddard's boyfriend. I just heard the news on the radio. Can we meet?"

"Sure," I said. "I'm right here at the office."

There was a heavy silence on the line. "I'm afraid, Mr. Faraday. I daren't come there."

"You name the place," I told him. "Where have you been hiding?"

"We lit out for San Francisco after Lisa took the stone. I told her no good would come of it and we had no means of selling it. Unless, of course, she was acting on someone else's instructions. But she told me it was just an impulse thing."

"An expensive impulse," I said.

"I still can't believe it," he went on. "We only got back in town this morning."

"That was a crazy thing to do as the heat may have been on."

"It's my guess she'd hidden the diamond somewhere in her place," Romano continued. "And she wanted to pick it up. But she was pretty close-mouthed about the whole business, though I questioned her many times."

I stared at the ceiling in disbelief. Amateurs. But I didn't tell him that. Then he gave me a time and place where we could meet and I said I'd be there in half an hour.

III

It as a little bar called Luigi's, in a discreet location off the main boulevards. Romano turned out to be a sallow-faced character of about forty, with a sweaty forehead that went with his agitated manner. I ordered a white wine and joined him the secluded booth in a dark corner. There weren't many people in and a radio was playing a number by Guy Lombardo. It was that sort of place. Romano was a frightened man all right and he didn't say a word after he'd identified himself, until the waitress had brought the order and he'd replenished his own glass.

"You'd better talk fast," I said. "The law takes a dim view of withholding evidence.

"I know, Mr. Faraday. I hope you'll keep me out of it. I'm talking to you because of Melissa. I owe it to her memory."

"Very touching," I said. "But we're wasting time. You may be in danger yourself."

The sallow man licked his lips again. It was the fourth time since I'd sat down and it was beginning to get irritating.

"I'm going to ground tonight. With my sister. She lives a thousand miles from here. I won't come back for a long time."

I stared at him. "Why the hell did you come back at all?"

He swallowed and took another pull at his glass. "Because of Melissa. And also I had to pick up some luggage and money from a friend's place. I needed to do that to get to my sister's."

"So what have you got to tell me?"

Romano drained his glass, his mind evidently made up. "We came back this morning," like I said. Melissa had some things to do here in L.A. What I wanted to tell you, Mr. Faraday, was that I think Melissa was working for someone else. Someone big. She didn't say much, but she mentioned Victor Zefferelli several times when we were on the lam. I believe she'd worked as a hat-check girl at one of his clubs a while ago."

I gave him one of my most impressive looks. "The big-shot gambler?"

"That's the one, Mr. Faraday. If it's any help . . . That's about all I know, but I wanted to do something for Melissa."

"You've been helpful," I said.

He put his hand across the table. "Thanks, Mr Faraday. I won't forget you."

"I'll try to forget you," I said. "Let's hope the police do the same."

I left him without a backward glance, got in the Buick and started making time. I rang Stella from a booth and gave her the score and then tooled on over to my house on Park West, freshened up, made myself a sandwich and broke out the Smith-Wesson .38, loaded it and was just putting my feet up on the kitchen table when the phone buzzed. I was getting really popular today.

"Mr. Faraday?"

The voice was soft and cultured. I admitted it and the caller went on. "Victor Zefferelli here. I think we have some mutual business to discuss."

I frowned at the ceiling. I was getting good at that. "Word certainly gets around. This wouldn't be in the area of precious stones?"

"Could be," he said cautiously. "The grapevine has been busy with Van Boren's loss and your name came up."

"Where you at?"

"My club, the Miramar. You know how to get here?"

"Sure," I said. "Been fleeced there many times."

He gave a harsh laugh. "Not in my day, Mr. Faraday. We run a straight business."

"If you say so," I told him. "Be there within the hour."

I locked up and with the Smith-Wesson making a comforting pressure against my chest muscles I put my size nines on the cement path that edges the garden. Maybe I was getting a little careless and perhaps distracted by me headlights of a passing sedan on the boulevard, but I suddenly heard a soft hissing noise that had no business to be there in the relative calm of Park West. I hit dirt instinctively and something chopped a branch off a bush about eight inches from my head.

I was hugging the earth and I had the Smith-Wesson up and ready now. Then I spotted the big Caddy in the shadow of the trees opposite. I fired once in the air, the explosion sounding thunderous. The auto gunned up with a screaming of tires and I just had time to get the California licence plate. Then I got up and dusted myself down. You're getting somewhere, Mike, I told myself. So long as it wasn't a mortuary slab. Nevertheless I gave myself a crooked smile in the driving mirror all the way across town.

The Miramar was as plushy as they come, but I had no time for the decor tonight. I parked the Buick next to a row of vehicles and went down the line. There was a big Caddy parked at the far end. I checked the number-plate and it fitted. I had a look at the licence details strapped round the steering post. It was registered to someone called Conrad Geiger. I broke out the Smith-Wesson and put the remaining shells in my pocket as well as the spent one. I went on into the Miramar and drifted up on to the mezzanine level. There I caught the eye of a bullet-headed barman called Tim Mabey whom I happened to know.

"I have an appointment with Zefferelli," I told him.

He jerked his thumb to the left where a velvet drape hung beneath an illuminated sign which said PRIVATE.

"Sure, Mike. Just go on through."

I eased back the curtain, opened the door and found myself in a long corridor with dim lighting and some tasteful nudes from the floorshow framed along the walls. I followed the arrows up to the far end where a gorilla with wall-to-wall shoulders was sitting at a table with a whiskey glass at his elbow and a copy of the *Examiner* spread out in front of him. He wore a white tuxedo with bulges that weren't entirely due to muscles and he stared at me with eyes like two cesspits. I figured him for quite an amiable person.

"I want to see Candy Barr," I said.

She was Zefferelli's secretary and for some reason had taken the name of a famous film actress I'd once worked on a case for. Perhaps she didn't want her friends to know of the Zefferelli connection. I'd known her from way back, and I always wondered why she worked for such a character, but she said the money was good.

"Who wants to see her?"

"I'm her brother Zanzi," I said.

The corners of his mouth flowered in a smile about one millimetre wide. "I make the jokes around here," he said.

"House rule?" I asked. "I didn't know. You must be a barrel of laughs."

He got up from the chair. He was about eight feet tall and broad with it. He looked me over again.

"You a private dick?"

I admitted it. "Faraday. Zefferelli wants to see me."

He grunted. "We got some classy-looking broads checked in here. We have to be careful."

"I can imagine," I said.

He opened his mouth another three millimetres. "I'll take the cannon," he said, reaching for my holster. He took the Smith-Wesson out gingerly and I broke it open for him. His eyes widened as he glanced at the empty chambers. A baffled expression rippled across his wooden features. He sniffed at the weapon suspiciously. "It's been fired recently."

"I like ferret shooting at night," I said.

He clamped his mouth into an even tighter slit and reluctantly handed the revolver back to me.

"Been out tonight?" I said. "It is Mr. Geiger isn't it?"

He changed gear on some gum in a corner of his mouth. "Maybe," he said, his eyes narrowed to tiny points. He had a great sense of humour. But he got the message. He reached for a phone on the table. "Round the corner, first door on the right."

I thanked him politely and went on down. When I got beyond, the angle of the wall I reloaded the Smith-Wesson and put it back in the holster. Then I rapped at the first door.

"Come," a girl's voice called.

I went on in.

IV

Candy Barr was as sweet as her name implied. She was a girl of about twenty-five and she came across the outer office from her desk with a broad smile of welcome. She was all curves and as she reached up to kiss me on both cheeks I felt the shock-waves reaching down to my socks and curling my toes. She was that sort of girl and had almost as powerful reaction on me as Stella. Today she wore a smart blue tailored business suit with an open-neck shirt which set off her fine neck muscles. Sue must have known what I was thinking because her smile broadened, revealing dazzling white teeth.

"It's been a long time, Mike."

"Too long," I said.

"What is that supposed to mean?" she threw back over her shoulder as she led the way back to her desk.

I grinned. "Anything or nothing. You haven't changed."

She sat down facing me on the front side of her desk, showing lots of leg, which shot my morale to hell and back.

"But you have, Mike."

"Whatever that means."

She shook the dark ringlets of her hair. "You used to be much more ardent."

"I've grown up since then."

She made an elegant little gurgling noise in the back of her throat. "That will be the day."

Just then the phone on her desk buzzed, bringing this promising conversation to an end. I wondered, for the thousandth time why she had chosen to be the front girl for a dubious character like Zefferelli, because she was as straight as they come. I figured it might have been because she had an incapacitated mother who needed constant expensive medical treatment. It was none of my business anyway. A muffled dialogue from the other end of the pnone was punctuated by Candy Barr's soft monosyllabic replies. She put the receiver down with a regretful smile.

"He knows you're here, Mike. You'd better go on in."

I bent over to brush her forehead with my lips. "We'll take this up some other time," I said.

I could feel those brilliant green eyes burning holes in my back as I went across the parquet toward Zefferelli's inner office.

Zefferelli, when I got to see him, was a smallish man ensconced behind a desk only slightly smaller than the reception area of LaGuardia Airport. He gave me an affable smile as I flogged toward him, exposing a nice set of gold-capped molars. A TV set near him was showing an old Warner Bros. gangster movie. Very apposite, I thought, but he switched it off with a regretful expression as I got up closer.

"Sit down and have a drink, Mr. Faraday. Bourbon? Best bonded stock."

I nodded, sinking into the leather chair that was strategically lowered so that he could dominate whoever was sitting in it. I guessed he'd been seeing too many old newsreels of the Third Reich. I glanced at the tumbler he'd slid across to me.

"The bottom of the glass is damp," I said.

He gave me another view of his gold-plated teeth but his smile was twisted this time. "That's the drink. This stuff is expensive. But we're not here to discuss the finer points of wines and spirits."

"I'm lucky to be here at all," I said. "Someone took a shot at me as I was on my way over."

Zefferella made an irritated clicking noise with his tongue. "That's too bad,

Mr. Faraday. But then L.A. is a dangerous place." He spread his hands wide on his blotter.

"You don't seems surprised," I said. "But the monkey outside was at the other end of the silencer. I got his number plate as he was making himself scarce."

The man behind the desk made another clicking noise. "That's terrible, Mr. Faraday. I shall have to speak severely to him. But let's forget that little episode and get down to facts."

"As you said on the phone, we're talking about the same business."

"A fortune, Mr. Faraday. In which I have a vested interest, you might say."

Zefferelli leaned forward across the desk and distorted his face into what I took to be an amiable expression. Before he could speak I cut in.

"You know who's got it now?"

He shrugged. "A diamond . . ."

"As big as the Ritz," I finished for him.

He got the reference. I was surprised he was so erudite. He obviously knew all about F. Scott Fitz.

The small man in the dark blue tailored suit that was intended to mark him as an honest businessman was intently studying his well-manicured nails. "Let me give you some good advice, Mr. Faraday. Leave this business alone. It's not worth the candle. You might have enquired into the provenance of this stone. Or all the others your client owns, come to that. I've done business with Van Doren so I know of what I speak."

"You're remarkably well informed, Mr. Zefferelli."

He shrugged again. "That's my business."

I took a cautious sip at my glass. It was pretty good stuff at that. "So he has no title to the thing?" I asked.

He nodded. "Exactly, Mr. Faraday. Despite what I've said, we could work together on this."

I shook my head. "I don't think so, Mr. Zeferelli."

He spread his hands wide on the desk again. "You're not averse to money, I take it?"

"Not at all. It's where it comes from that bothers me. I like to sleep nights."

He gave his teeth another airing. "But there's no need for us to be enemies, then?"

"You'd better tell your gorilla that," I said. "Next time I'll put a hole in him."

He said nothing but sat with an impassive face. I thanked him politely for the drink and went on out.

V

"Don't forget the autopsy," Stella said.

I was resting my arches on what was left of the office carpet and had been watching a couple of bluebottles executing a graceful *pas de deux* against the window blinds. I must have dropped off, salivating like Pavlov's dogs at the thought of the brew Stella was about to concoct. It was ten of eleven and my mind was on other things.

"What autopsy?" I asked.

Stella gave me a withering look from under the gold bell of her hair. "The Melissa Goddard number, goof. There was a piece in the *Examiner* this morning."

"Oh, that," I said.

Stella gave me another steely glance. "I thought you were on the case," she said. "Van Doren's been on three times already. I told him you were hot on the trail."

"Good girl."

I heaved myself up in my swivel chair and tried to look intelligent.

"There's a character called Dr. Abraham Sorensen," Stella went on. "He's the pathologist on the case. He was due to perform the autopsy this morning. He might come up with something interesting, so you'd better drag your underwear over to the City Morgue."

I gave her one of my appealing little boy looks. "After the coffee, honey," I said.

Dr. Sorensen, when I got to see him, was a small, balding man of indeterminate age with a worried expression and shifty eyes. I didn't take to him. But then I didn't have to. The Morgue was a vast, cheerless building which smelt of antiseptics and carbolic soap, but Sorensen's office was a lot better. For example he had a Swedish secretary called Inga Svensen, who came at me in curves when I opened the door. We were having an interesting conversation, with our heads close together when Sorensen erupted into the room, giving an exasperated snort at the same time.

"What's all this?" he snapped, swivelling his watery grey eyes from behind the gold-rimmed pince-nez, first at the girl and then at me.

"She's my cousin," I said. "We were just reminiscing about the old days in Stockholm."

The tall blonde number gurgled and moved away to her desk. A bewildered look passed across Sorensen's unattractive face. I guessed he wasn't used to repartee in his line of business. His clients never answered back. He put down a clipboard he was carrying on the blonde number's desk and barked out a few instructions. He wore a white surgical smock which didn't make him seem any more attractive.

"I suppose you want to see me?" he snapped.

"I'd prefer Inga," I said. "But you'll have to do."

He stared at me with something like suppressed fury and then grunted, "Come into my office."

We went on into his small, austere room, which was fitted up with a grey steel desk, a few chairs and four grey metal filing cabinets. Green shaded lamps hung from the ceiling which had seen better days. He didn't ask me to sit down but I sat down anyway, sliding the photostat of my licence across his desk. His eyes widened and a muscle fretted in his left cheek.

"Oh," he said. "So you're the private dick who found the Goddard girl's body."

"That's right," I said. "I'd like to get a look at your report. I understand you carried out the autopsy earlier today."

He started back as if I had stung him. "That's strictly private," he snapped, little red spots standing out on his cheeks.

"I think not," I said. "I have Lieutenant Mulcahy's permission. You can ring him at the precinct if you like."

I pointed toward his phone. He licked suddenly dry lips.

"That's different," he said. "Why didn't you say so."

"You didn't give me a chance," I told him.

He rummaged around on the top of the desk. "I carried out three this morning," he said in more placatory tones. My secretary hasn't typed up my notes yet."

"That's okay," I said. "I just wanted a quick skim through."

"Perhaps you'd like to see the body," he added with a slight sneer.

I shook my head. "I saw it while it was still warm. I'd rather not view it after you've hacked it about."

There was no need for me to say that, but he had me riled up with his attitude. I had a sudden thought.

"I suppose she didn't swallow anything."

Sorensen flushed and gave a short barking laugh. "That diamond was so big it would have choked her." His eyes were narrowed now and he was really mad. "I read the papers, Mr. Faraday. Mr. Van Doren was round here this morning too."

"The hell he was," I said.

"You're working for him, aren't you?"

I stared him down. "These things are supposed to be confidential. But if Van Doren has been shooting his mouth off to you, I can't very well deny it."

"That's better, Mr. Faraday. Now we know where we stand."

"Wherever that is," I said.

Sorensen got up, handing me several scribbled sheets of notepaper. "You'd better read this anyway. It will give you all you want."

Now that he was up and standing in front of me I saw that he was wearing shoes with Cuban heels to make him look taller. They were old and cracked at that. He saw where I was looking and flushed.

"It gives me a slight advantage when dealing with much taller people," he said defensively.

"Sure," I said. "Alive or dead."

A slight flicker of a smile passed across his unattractive features. It didn't do anything for him. "Alive, of course. Once on my slab I have the advantage."

I didn't bother to answer that but settled down to read his report as he went on out to consult with the blonde number.

VI

Two days had gone by and nothing much had happened. Unless you count half-a-dozen frantic calls from Van Doren and an interview downtown with one of the D.A.'s aides, a frosty-faced character called McGinty, who did everything but call me a liar and obviously suspected me of the Goddard girl kill.

But Mulcahy and I had a good laugh afterwards.

"He's a nut case," he'd said. "He won't last long in the D.A.'s office at that rate. The D.A. had already cleared you in any case. The guy just loves throwing his weight about. I suppose he feels he has to justify his salary."

"Don't we all," I'd said. "If I had a salary, that is."

Now I was polishing the seat of my pants in my swivel chair in the office and counting the stalled cars on the boulevard below. Just then Stella came in so I made like I was thinking, but it didn't fool her. She was carrying a copy of the *Examiner* and slapped it down on my desk.

"Take a look at that, Mike."

"Very nice," I said.

Stella sighed. "Not the bathing beauty parade, dumbo. The stop press."

I blinked and sat upright.

It was a piece about a corpse found in a blazing sedan in a ravine not a dozen miles from where we were sitting. It was the plug-ugly who'd tried to ice me a few nights ago. I wouldn't have to watch my back any more.

Stella's eyes were sparkling as she went across to brew up. I fished in my drawer for the small notebook which contains phone numbers I'm not supposed to have. I soon found what I wanted. It was a direct line to the guy's private office. A familiar voice came on the line.

"Yes?"

"Many thanks, Zefferelli."

"Who the hell is this?"

"Your old pal, Faraday."

"How did you get this number?"

"I've had it for years," I told him.

That didn't go down too well as I could gather from the heavy silence.

"Whaddya want?"

He was forgetting his diction now.

"Just thanking you for getting rid of Curlytop."

"I don't know what you're talking about."

"You should read the papers. And thanks anyway," I added and put the phone down.

Stella placed the steaming cup down on my blotter and skipped out of range with the skill born of long experience. The day was beginning to shape up nicely. I felt it was time I spoke to Sorensen again so I left the details with Stella in case anyone but Van Doren needed me, and tooled over to the Morgue. The blonde number was still in the office and seemed inclined to fraternise, but it was too hot for that sort of thing today. I was taking my size nines off to the private office when she stopped me.

"He's not in today, Mr. Faraday. But I can give you his private address if it's urgent."

"There are a few points I'd like to discuss," I said. "But don't tell him I'm coming."

She exposed her beautiful set of teeth once more and scribbled something on a sheet of headed notepaper. "I understand."

Her smile lasted me all the way to the Buick.

VII

Sorensen's place was a large bungalow spread on a box canyon way across town and I got up there in the early afternoon. It was a pretty lavish set-up with lots of tropical plants and manicured lawns, but I had no time for the gardening notes. There was a big black sedan parked out front and I eased up the cement path slowly, keeping in the shadow of the overhanging trees. I felt the radiator grille; it was cold and I saw the licence strapped round the steering post was made out to Sorensen. So he had to be home. But I've kept on breathing in the rough trade I've chosen by being cautious, so I didn't press the buzzer in the front porch.

Instead I tried the big oak door. Like I figured it was locked. So I went on around in back, treading lightly on the balls of my feet, ducking below window level. There was an emperor-sized pool with a springboard and the yellow-green water was sparkling in the sun and looked very inviting, but I wasn't here for the swimming either.

I half expected to see Sorensen in the pool; face down and floating limply. It was that sort of case. Besides, I'd seen *Sunset Blvd.* again on TV a few nights ago. But I smelt trouble when I found the back door unlocked. I went on into the shady interior, easing out the Smith-Wesson from the shoulder holster.

There was an odd atmosphere in the place. Then I caught it; the unmistakable acrid tang of powder smoke. I was through into a big room that had an enormous stone fireplace and framed diplomas on the walls. Then I saw the reason for the atmosphere. Sorensen had carried out his last autopsy. He was sprawled on a white leather divan that was now mottled red. Someone had done a nice job of icing him. There was a small bullet hole over his right ear, with scorch marks around it. Whoever had done this had to be pretty intimate to get that close.

The mess was all on the other side so I didn't turn him over. I'd seen too many in my profession. I smelt the faint odour of a woman's perfume then. Ladies always use a small calibre automatic in my experience. The gorilla who'd shot at me was dead too, so that ruled him out. It left only the lady in the case because the kill didn't carry the stamp of Zefferelli's boys. And, like the Goddard number's place, someone had already been over the house in a very amateurish way. Even so I added to the chaos though without much hope of finding anything. I put the Smith-Wesson back in the holster, where it rested comfortably against my chest muscles.

I went quietly up the stairs and through the bedrooms. Some of them looked as though they'd been used by the lady in the case. And I certainly didn't think the cleaning lady wore black silk knickers. The perfume was coming from that drawer and while I was still thinking about it a shadow passed across my field of vision. I looked into Inga Svensen's dark green eyes and from them to the little automatic she held steadily on my gut.

"You got here quickly," I said.

A smile briefly flickered across that innocent-looking face. "I tried to warn him you were coming, but he'd cut the phone off. I know a short cut out here that takes several miles off the route. Sure, Abraham must have found the diamond all right. We planned to split fifty-fifty. I figured he was trying to skip out and leave me behind. I found him packing when I got here half-an-hour ahead of you."

"And sent him packing," I said. "The perfect secretary."

She shrugged. "You don't know what it's like being tied up with a guy like that."

"I can guess," I told her. "But my tastes don't lie in that direction."

Again the faint flicker of a smile. The gun muzzle was still trained steadily on my belly. She wrinkled up her brow.

"Where do we go from here?" she said softly.

"The gaol's the next stop," I said as a car gunned up outside.

"Come in, lieutenant," I called.

The girl turned her head briefly as my balled fist caught her on the side of the jaw. She went over backwards, hitting her head on a corner of the dressing table, while the automatic went skittering over the carpet. I put my foot on it and broke out the shells. Like I figured, only one shot had been fired. I wrapped the weapon in my handkerchief in order to preserve the prints, if any. The law is rather particular about things like that. Then I got to the girl; she had a cut on her forehead, which was bleeding slightly, but she'd live. She was a big girl, and I had a job to carry her downstairs, where I made her comfortable in another room which was got up as a study. Then I phoned Stella and told her what had happened. I could almost see her eyes widen at the other end of the wire.

"What a heel, Mike," she breathed. She seemed more concerned about the old man double-crossing the girl than her taking Sorensen out. But I could never figure out women's thought processes anyway. Then the idea struck me.

"I think you've got it, honey," I said. "Ring you later."

I made sure the girl was still out and then went quickly back into the living room. The cracks in Sorensen's Cuban heels showed up beautifully in the afternoon sun. I got a paper knife from his desk and inserted it into the crack on his left shoe. Nothing. I struck gold the second time around. The crack in the right heel widened as I applied pressure on the knife. The flap came back on a cleverly engineered hinge. It was an old trick, but it always worked. Inside the shallow cavity was a small twist of tissue paper. I held it up to the light and unwrapped it. I recognised the contents then. It was a key to a locker in a downtown bus station unless I missed my guess. The girl was still out, so I picked up the phone and got some law in.

VIII

"Where do you keep digging up these stiffs?" Mulcahy growled.

"I like doing your work for you," I said.

He gave me what passed for a smile in police circles. The meat wagon had been and gone. I'd given the law the girl's automatic and now I handed Mulcahy the key. I'd waited until the girl had been taken away in a police car with her still only half-conscious. I didn't want to be sadistic on such a nice afternoon.

"You know what you'll find there?" I said. "It was in the old guy's shoe."

Mulcahy's eyes glittered as he handled the key, which sparkled in the sunlight. "You sure you weren't going to use this yourself?" he asked, only half-joking.

I shook my head. "Amsterdam's too far and I couldn't get it through customs."

Mulcahy gave a short laugh and splashed some more of Sorensen's bourbon, making the ice cubes tinkle in the frosted glass. He puckered up his mouth in an expression of satisfaction as he lifted his own tumbler.

"And don't say this is the stuff that dreams are made of, Mike. I saw the movie."

"You and fifty million others," I said. "We'd better get down to Headquarters. You can close up the Zefferelli angle and I guess you'll want to know how Van Doren came by all his rocks."

He puckered up his lips again. "Just why are you telling me all this?"

"Because I'm an honest man," I told him. "And because I got paid up front."

His grin widened. He tipped his glass. "I'll drink to that."

IX

"You ought to get a medal, Mike," Stella said.

I smiled. Even the sheeting rain at the windows couldn't diminish the golden beauty of Stella's hair. You're getting one of your Emily Dickinson moods again, Mike, I told myself. But I had a point.

"We got something better than that," I told her.

She gave me one of her unnerving looks from very blue eyes. “And what might that be?”

“No might be about it, honey,” I said. “Van Doren stopped our bank balance from nose-diving into the underground car-park.”

“There is that,” she said. “Coffee?”

“Try me,” I said.

She went over to the alcove where we do the brewing-up and snicked the electric kettle on. Then she came back, sat down the other side of my desk, in the clients’ chair. When we have any, that is. She smoothed out her copy of the *Examiner* which carried a lurid spread with photogravure on the Goddard kill, together with shots of Van Doren, Zefferelli and the Svensen number. Not forgetting a long shot of Geiger’s burnt-out car and an appropriately smiling shot of Sorensen.

Stella turned the page and looked at another full-face of Inga Svensen and a couple of shots of the enormous rock all the fuss was about.

“One of your busiest cases, Mike,” she said absently.

“You could say that.”

She shot me a quick smile and then turned back to the pictures of the diamond.

Like always, the Homicide boys took all the attention. There wasn’t a mention of me, which suited just fine.

Stella gave a little sigh.

“What is it?” I said.

She still had her eyes fixed on the *Examiner* before her. “What a beautiful diamond. It would look wonderful round my neck.”

“Your neck looks great without all that lethal junk,” I told her.

I sat on for a moment, watching the faint flush that spread across her cheeks. Then I gave her a wry smile.

“Diamonds are for never,” I said.

E. OTHER BOOKS

Includes an alphabetical listing of other books written by Basil Copper, but not necessarily credited to him

(*"People sometimes ask me how to become a writer. Frankly, I don't know. In my opinion writers are born, not made. You can, it is true, train to be a journalist, but factual reporting is very different from being an imaginative writer. There are such things as schools and correspondence courses for writers, but although they can teach a person how to spell and punctuate, to lay out their typescript properly and other technical matters, if that person hasn't got the ability or imagination then it is ultimately useless. It is a peculiar process, the business of creative. One just sits down at a typewriter and the words immediately start to flow on to the paper. I write on a very heavy Olympia Standard typewriter–no electrics for me–and I pound away happily in the way I have always done from my days as a very young journalist. This suits me perfectly, but it would not do so for others, more attuned to word-processors and electronic machines."* BC.)

E1 **HOLD THE PRESS!**
[Note: as by "Basil Baxter"]
Novel.

a. __________, Sevenoaks, Kent, 2004 (chapbook)
Privately printed in an edition of 50 copies / 100pp.
[Note: "Basil Baxter is the pseudonym of a well-known writer who has written many books in various genres, which includes crime and non-fiction. This is his only comic volume, based on his experiences as a journalist and editor over a long period of time though all the incidents are exaggerated and transmuted into fiction for greater effect. This edition is privately produced for friends and former colleagues."]

E2 **THE STORY OF THE PHANTOM 3: THE SCORPIA MENACE**
Novel.

a. __________, Avon Books/King Features Syndicate, New York, December 1972 (pb)
Price: 75¢ / 158pp / Cover art: G. Wilson.
[Note: "Lee Falk's Original Story adapted by Basil Copper." Title page and spine read *The Phantom and The Scorpia Menace.*]

E3 THE STORY OF THE PHANTOM 2: THE SLAVE MARKET OF MUCAR

[as by "Lee Falk"]

Novel.

a. __________, Avon Books/King Features Syndicate, New York, November 1972 (pb)
Price: 75¢ / 159pp / Cover art by G. Wilson.
[Note: "Lee Falk's Original Story."]

Above: "Basil Copper on Film" panel, October 30, 1993 World Fantasy Convention, Minneapolis, Minnisota.

Above and right: Book release party at Sherlock's Home Restaurant, Pub, and Brewery, Minnetonka, Minnesota, October 31, 1993. Basil Copper's first visit to America.

Actor Christopher Lee and Basil Copper at
The 1997 World Fantasy Convention, London, England

IN THE FOOTSTEPS OF DRACULA

[Note: The following piece was written for the sleeve notes of a 1974 LP entitled *Hammer Presents Dracula with Christopher Lee*.]

THE LEGEND OF THE VAMPIRE embodies one of the oldest fears of humankind. And from the mist-haunted regions of the past it has echoed down the centuries so that its curse still reverberates with menace in our own day.

The cruel history of the Eastern European states and particularly Transylvania, now part of modern Romania, where the myth has most potent force, was crystallised in the real-life character of Vlad Drakula, a military governor of Wallachia in the fifteenth century, noted for his deeds of barbarous cruelty.

In fact he was known as Vlad Tepes—"Vlad the Impaler"—and ruled as a sovereign from 1456-1462, when he became a national hero in his wars against the Ottomans. But some of his habits were less agreeable; he impaled his enemies alive on spears and once, when a feudal lord was invited to one of his feasts, the guest complained of the smell of the corpses lining the room and was immediately impaled himself for his impertinence!

He was a master of ritual torture and ruled as an absolute despot; as well as battling against the Ottomans when they invaded Wallacia, Vlad was prominent in a fratricidal war between the branches of his own family, which went on until

by Basil Copper

almost everyone was killed.

Cruel as he was, Vlad was not a vampire in the true sense. That distinction belonged to the terrible Countess Elizabeth Bathory, a Hungarian noblewoman born in 1560, whose orgies of blood-letting in her ancestral castle involved the murder of hundreds of young girls, whose blood she drank. Girls were kept chained in her dungeons, being "milked" of their blood as the awful Countess needed it.

She was brought to trial in 1611 after the authorities raided her castle during an orgy and was sentenced to life imprisonment, being walled up in her own bedroom, where she died four years later.

She was in the direct tradition of the terrible Gilles de Rais, a fifteenth century Marshall of France who fought alongside Joan of Arc. In private he was a pervert obsessed by blood-lust, murdering, torturing and drinking blood. The Black Baron was executed in 1440.

In our own day, medical vampirism is not unknown, and there have been at least three classic cases in the last hundred years, the most recent in the late 1940s.

But it was the grim reign of Vlad which attracted the attention of the novelist Bram Stoker, who based his 1897 novel on this sadistic monster. The work, *Dracula*, became world-famous, has sold millions of copies and has never been out of print since first publication. It became a cult book and even today, thanks to the cinema and television, the myth of the vampire–the undead dead who returns to feast on the blood of the living–looms larger than ever in the world's subconscious.

The theme was given added impetus, firstly by Max Schreck, a silent film actor who brought awesome authority to the horrifying Dracula-figure of Count Orlok, in the classic horror film, *Nosferatu*, directed by the great F.W. Murnau in 1922.

In 1931 came the first sound *Dracula* in the cinema, the Count being personified by the Hungarian actor Bela Lugosi, who had first played the part on the American stage, and who was to go on playing Dracula in films in one form or another for nearly thirty years.

The most famous exponent of Dracula for modern audiences is the distinguished actor Christopher Lee, who took over the role for Hammer Films in 1958, and has repeated it with brilliant success, in varying forms, in many films since. Now he lends his considerable talents to create a chilling essay in vampirism for aficionados in an evocative and brilliant LP recording by Hammer City Records, *Dracula*.

Once again the most celebrated and terrifying monster in the genre, Count Dracula walks the fog-shrouded countryside of Transylvania, while thunder rumbles in the distant hills and the mournful howl of a wolf strikes chill to the heart.

Basil Copper as Guest of Honour at a Dracula Society dinner in the mid-1980s with the group's secretary, Julia Kruk

It is dusk—time for the vampire to awake; a broken-down coach is in the Borgo Pass and a beautiful young girl-victim in distress. What more could the enthusiast for *Dracula* wish for?

With the aid of a full symphony orchestra playing the music of composer James Bernard; menacing sound effects created by producer Don Norman; and an ingenious script by Don Houghton, Mr. Lee relates an entirely new adventure of a fiend who has horrified and delighted whole generations over the years.

Side Two of this vivid and unique recording features the fine atmospheric music for four Hammer productions. *Four Faces of Evil* includes the first-rate mood music arranged and conducted by Philip Martell from *Fear in the Night*; *She*; *The Vampire Lovers* and *Dr. Jekyll and Sister Hyde*.

I am sure that the listener, safe and secure in his own home, cannot but agree with Christopher Lee and the creators of this brilliant recording—DRACULA LIVES!

F. NON-FICTION

Includes an alphabetical listing of non-fiction works written by Basil Copper

F1 **AUGUST DERLETH: A GIANT REMEMBERED**

a. __________, in *August Derleth Society Newsletter*, Vol.1, No.2 (1978).

b. __________, see this volume.

F2 **AUTHOR'S NOTE**

a. __________, see Necropolis [A6a].

b. __________, see Necropolis [A6b].

F3 **AUTHOR'S NOTE**

a. __________, see Solar Pons versus The Devil's Claw [C10].

F4 **DARKNESS, MIST AND SHADOW—THE LANDSCAPE OF FEAR**

a. __________, in *Dark Horizons* 16, Spring 1977, edited by Geoffrey Noel Smith, British Fantasy Society, Whitstable, England (magazine).

b. __________, see this volume.

[Note: 'Guest of Honour's Lecture' presented at Fantasycon 3, Birmingham, England, on February 26, 1977.]

F5 **ECSTASY AND TIME TRAVEL IN SEVENOAKS: THE GENESIS OF MODERN SCIENCE FICTION**

a. __________, in *The H.G. Wells Scrapbook: Articles, Essays, Letters, Anecdotes, Illustrations, Photographs and Memorabilia About the Prophetic Genius of the Twentieth Century* edited by Peter Haining,
New English Library, London, England, 1978 (hc).

b. __________, in *The H.G. Wells Scrapbook: Articles, Essays, Letters, Anecdotes, Illustrations, Photographs and Memorabilia About the Prophetic Genius of the Twentieth Century* edited by Peter Haining,
Clarkson N. Potter, New York, New York, 1978 (hc)*.

F6 **FOREWORD**

a. __________, in *The China Roundabout* by Josephine Bell, Chivers Press/ Black Dagger Crime, England, 1996 (hc).

F7 **FOREWORD**

a. __________, in *A Delivery of Furies* by Victor Canning, Chivers Press/ Black Dagger Crime, England, 1987 (hc).

F8 **FOREWORD**

a. __________, in *The Fingerprint* by Anthony Gilbert, Chivers Press/ Black Dagger Crime, England, 1991 (hc).

F9 **FOREWORD**

a. __________, see Not After Nightfall: Stories of the Strange and the Terrible [B6].

F10 **FOREWORD**

a. __________, in *The Saint Goes West* by Leslie Charteris, ChiversPress/ Black Dagger Crime, England, 1992 (hc).

F11 **FOREWORD**

a. __________, in *Snake in the Grass* by Anthony Gilbert, Chivers Press/ Black Dagger Crime, England, 1993 (hc).

F12 **FOREWORD**

a. __________, see Solar Pons The Final Cases [C9].

F13 **FOREWORD**

a. __________, in *The XYY Man* by Kenneth Royce, Chivers Press/ Black Dagger Crime, England, 1995 (hc).

F14 **THE GAME'S AFOOT: AUGUST DERLETH AND SOLAR PONS**

a. __________, in *Return to Derleth: Selected Essays Volume Two* edited by James P. Roberts, White Hawk Press, Madison, Wisconsin, August 1995 (chapbook).

[Note: "Due to a printing error, pages 10 and 83 are transposed."]

b. __________, see this volume.

F15 THE HAUNTED HOTEL

a. __________, in *Dancing With the Dark: True Encounters With the Paranormal by Masters of the Macabre* edited by Stephen Jones, Vista/Cassell, London, England, 1997 (pb).

b. [as **STRASIDELNY HOTEL**], in *Valcík s Temnotou: Mistri Hororu Vyprávejí o Svych Setkáních se Záhadnym a Nevysvetlitelnym* edited by Stephen Jones, Apsida-Knizni Klub, Praha, Czechoslovakia, 1998 (hc).

c. __________, in *Dancing With the Dark: True Encounters With the Paranormal by Masters of the Macabre* edited by Stephen Jones, Carroll & Graf, New York, New York, 1999 (tp).

d. [as **L'HOTEL INFESTATO**], in *La Danza delle Tenebre* edited by Stephen Jones, Bompiani, Milan, Italy, 1999 (tp).

(*"No, this has nothing to do with Wilkie Collins' famous macabre tale, but relates to a series of ghostly happenings experienced by myself and friends. But I should perhaps first state that I am not, for various reasons, going to give the actual locations and names of persons involved, other than myself. I hasten to add that this is not because the events related are fictitious–far from it–but because the owners of the hotel would hardly be pleased at such publicity, nor the present owners of the house concerned. In fact I was a young journalist when these incidents took place and actually wanted to write a series of articles on the happenings, but the hotel chain issued a stern veto with which my editor concurred."* BC.)

F16 THE HORROR HORN: THE BEST HORROR STORIES OF E.F. BENSON

a. __________, in *Horror: 100 Best Books* edited by Stephen Jones and Kim Newman, Xanadu Publications, London, England, 1988. There are two variants of this edition.

i. Description: trade edition (hc).

ii. Description: 300-copy signed edition (hc).

b. __________, in *Horror: 100 Best Books* edited by Stephen Jones and Kim Newman, Carroll & Graf, New York, New York, 1988 (hc).

c. __________, in *Horror: 100 Best Books* edited by Stephen Jones and Kim Newman, Carroll & Graf, New York, New York, 1900 (tp).

d. __________, in *Horror: 100 Best Books* edited by Stephen Jones and Kim Newman, New English Library, London, England, 1992 (tp).

e. __________, in *Horror: The 100 Best Books* edited by Stephen Jones and Kim Newman, Carroll & Graf, New York, New York, 1998 (tp).

(*"In describing some of my favourites, I hope to strike a sympathetic chord in the reader and can only hope that she or he will turn to some of Benson's neglected masterpieces, whose very pages seem to breathe out the odour of decay. But there are also many delights among the horrors."* BC.)

F17 **IN THE FOOTSTEPS OF DRACULA**

a. __________, on *Hammer Presents Dracula with Christopher Lee*, Hammer City Records/EMI Records Ltd., Hayes, Middlesex, England, 1974. (TWOA 5001). Side One: 'Dracula': Narrated by Christopher Lee, Introduced by Bill Mitchell, Script by Don Houghton; a. *Scars of Dracula* (James Bernard), b. *Taste the Blood of Dracula* (James Bernard). Side Two: 'Four Faces of Evil' The Hammer City Orchestra: 1. *Fear in the Night* (John McCabe), 2. *She* (James Bernard), 3. *The Vampire Lovers* (Harry Robinson), 4. *Dr. Jekyll and Sister Hyde* (David Whitaker). Music arranged and conducted by Philip Martell, Engineer: Bob Auger, Produced by Don Norman.

b. __________, on *Hammer Present Dracula with Christopher Lee*, Hammer City Records/Capitol Records, Inc., New York, New York, USA, 1974. (ST 511340).

[Note: "A horrifying story of vampirism with spine-chilling sounds and music." Basil Copper wrote the sleeve notes for this 33 r.p.m. vinyl LP recording of a new Dracula story narrated by Christopher Lee plus music from various Hammer films.]

c. __________, see this volume.

F18 **INTRODUCTION**

a. __________, see The House of the Wolf [A4b].

F19 **REMEMBERING AUGUST DERLETH AND ARKHAM HOUSE**

a. __________, in *The Architecture of Fantasy & Horror: World Fantasy Convention, Madison—2005* edited by Richard Gilliam, World Fantasy Convention, Madison, Wisconsin, November 2005 (convention souvenir book)

b. __________, see this volume.

F20 **THE VAMPIRE: IN LEGEND, FACT AND ART**

a. _________, Robert Hale & Company, London, England, 1973 (hc)
Price: £2.50 / 208pp.
[Note: A 1972 uncorrected proof edition of this book also exists.]
b. _________, The Citadel Press/Lyle Stuart, Inc., Secaucus, New Jersey, June 1974 (hc).
Price: $6.95 / 208pp / Jacket design by Judith Kazdym Leeds.
[Note: The title on the dust-jacket is *The Vampire in Legend and Fact.*]
c. [as **DER VAMPIR IN LEGENDE, KUNST UND WIRKLICHKEIT**], Wilhelm Goldman Verlag, Munich, West Germany, 1974 (hc)
Price: €18.90 / 186pp.
d. _________, Corgi Books, London, England, 1975 (pb)
Price: 50p / 222pp.
e. _________, The Citadel Press, New York, New York, June 1989 (tp)
Price: $7.95 / 208pp.
[Note: The title on the dust-jacket is *The Vampire in Legend and Fact.*]
f. _________, Robert Hale Limited, London, England, 1990 (tp)
Price: £5.95 / 208pp.
g. [as **DER VAMPIR IN LEGENDE, KUNST UND WIRKLICHKEIT**], Festa-Verlag, Leipzig, Germany, June 2005 (tp)
Price: €12.90 / 335pp.

("The Vampire *is intended to appeal to those people who would like as much material as possible on the subject within two covers; to those whose appetite–literary, I hasten to add–might be whetted further and thus be persuaded to turn to such magnificent stories as 'For the Blood is the Life' and 'Mrs. Amworth'; and finally to those who, like myself, have an insatiable interest in the macabre.*

"I'm afraid this is a very flawed work, due to a number of circumstances: firstly, as I started nearly three years' work on it in 1970 it was impossible to see copies of films like Bela Lugosi's Dracula. *I saw it when I was very young (a friend who was the cinema manager let me in) and I remembered the line 'Children of the Night; what music they make!' as 'Listen to them howl'.*

"Also, a friend in Hollywood told me that in the original film the Count was seen descending vertically down the wall of his castle, with his cloak spread out around him, but it was cut out of the release print. I repeated this, but it was completely untrue.

"Nobody seemed to notice, and when the book was reissued as a large paperback by Hale in 1990 I did a great many corrections." BC.)

F21 THE WEREWOLF: IN LEGEND, FACT AND ART

a. __________, Robert Hale, London, England, 1977 (hc)
Price: £3.95 / 240pp / Cover design by Colin Andrews.
b. __________, St. Martin's Press, New York, New York, 1977 (hc)
Price: $8.95 / pp 240 / Cover design by Colin Andrews.
[Note: A proposed third volume in this series, *The Monster: in Legend, Fact and Art*, was never written.]
(*"The legend of the werewolf has been known since Roman times and earlier; in our present age of space-travel and the atom bomb it is still a forceful and colourful figure. I have had a great deal of pleasure in tracing its genesis from earliest times down to the present day and in so doing I hope I have been able to communicate this to the reader."* BC.)

F22 WEREWOLVES IN LEGEND AND LORE

a. __________, in *The Scream Factory* Issue #15, Deadline Press, Los Gatos, California, Autumn 1994 (magazine).
[Note: "Special Werewolves Issue".]

Stephen Jones, Basil Copper amd Randy Broecker, January 1997, London, England

THE CURSE

[Note: Basil Copper's first published story, written for his school magazine when he was fourteen years old.]

I DON'T SUPPOSE you remember Colonel Gordon? Queer face he's got you know—egg-shaped, with two huge bushy eyebrows near the top; underneath this his eyes, large and piercing; then his fine curved nose with an atrocious little tooth-brush moustache directly below; and, lastly, his thin, bloodless lips with his strong craggy chin jutting put beneath.

A few months ago I bumped into him coming out of the Ovul Club.

"Ah, my dear Chalmers!" he said, on catching sight of me. "Come into the club-room. I have something extraordinary to tell you."

I followed him into the cosy club-room, where he drew up a chair for me by the fire. Having sent off the waiter for a bottle of Martini, the Colonel commenced, "You know old Forsythe, don't you?"

I nodded. The memory of him almost made me laugh. He was a huge great fat man, with little piggy eyes which were continually winking and blinking behind the huge horn-rimmed glasses he always wore.

"Well, remember his name won't you," said the Colonel, "for he has a great bearing on the story I am going to tell you.

"When we were stationed out in India, Forsythe fell in love with a girl and eventually became engaged to her. The wedding arrangements were almost

by Basil Copper

completed when some other fellow stepped in and carried her off. At this Forsythe went absolutely mad and threatened to kill him (the other fellow) and in the heat of his excitement he made a curse. He vowed that when he died the other fellow and the girl would shortly follow him. As they had crossed his path in life, so he would cross theirs in death. He added, why, I can't guess, that when he died he would visit me. I didn't know what that meant until this morning. You know many strange things come to pass in the East.

"Well, the other night, as I was going to bed, I felt vaguely disturbed. I hadn't been feeling well all day, but I put it down to the weather at the time. It was the same when I went to bed. The air was oppressive and stifling, and I couldn't get to sleep. At about half-past one in the morning I heard something that brought the sweat out on my forehead and made the blood run cold in my veins. It was a soft insistent tapping, intermingled with a voice muttering fearfully, 'Gordon, let me in. Let me in.' Then it died and then rose again, quivering on the midnight air, finally tailing away into an eerie silence. A long time after it had gone I sat there, sweating in fear, the blankets huddled round me. There was no more sleep for me that night, I can tell you.

"This morning I received a telegram from Australia, stating that Forsythe was dead. Then I remembered the curse. I had forgotten. But Forsythe didn't forget; he kept his word. Strange, wasn't it?

But, I say, look at the time. I must be off. Cheerio, old man."

After the Colonel had gone I sat looking into the fire. Last week my wife had died. I was to be the next. You see, I was 'the other fellow'.

PROFESSOR MAHMOUD

In memory of my Egyptian years.
—BC.

I

I FIRST MET Professor Mahmoud under the most mundane circumstances. I had just been crossing the seething turmoil of Ramley Square where the ancient trams which ran up from the docks lingered a while before travelling on to the suburbs of this great city. I had safely gained the Grand Corniche, the superb esplanade which the French had constructed in massive granite blocks and which runs for several miles in a huge arc round Stanley Bay, forming a rampart separating the city of Alexandria from the sea.

I was enjoying my walk in the eternal sunshine when I became aware of an elderly gentleman in front of me who seemed to be in great pain. He was elegantly dressed and had the look of a scholar about him with his thin, ascetic face and gold-rimmed pince-nez. I hastened to his side and asked if I could help. He was obviously of Arabic descent, but answered me in perfect English.

"You are very kind, sir."

He had apparently twisted his ankle by stumbling across the kerb and had fallen to the ground before picking himself up. I put my hand under his am and half-lifted him to the edge of the road, where I hailed a passing *gharry* as he told

by Basil Copper

me he detested motor vehicles and used them as little as possible. Then, despite his protests, I said I would accompany him to his home. And then, as the horse clip-clopped its way through the dense traffic along the Corniche, I learned something of his history.

He was a very friendly person and as our relationship developed, as I will relate in due course, I learned that he had been a professor at Cairo University and had a long and distinguished academic career. He knew London and Paris very well and, as he spoke perfect French as well as English and classical languages, he had lectured at the great seats of learning in England and at the Sorbonne.

He had a great sense of humour and had once observed to me in later years that Kafka was the poor man's optimist. He was referring to the wretched plight of the *fellaheen*, the abject poor of Egypt, not only in rural communities but in the great cities where they were reduced to begging in the streets. All this, of course, was long after I had formally introduced myself during that memorable *gharry* ride.

The Professor was a compassionate man and was very concerned over the dreadful conditions in which the poor existed. They began this wretched form of existence at a very early age as shoeshine boys pestering Europeans in the cities and covering their shoes with some dreadful slime in order to earn a few *piastres* in cleaning them.

When we eventually arrived at the elegant suburb of Sidi Bishr, which I had guessed to be our destination, the professor's home turned out to be a spacious French-style villa with immaculate gardens.

His gardener was just abort to leave but ran forward to assist me in carrying the Professor up the front steps. The interior was cool after the insistent heat of the sun, and with the very high ceilings all such houses have in the European quarters of Middle Eastern countries. His servant was out, but I made him comfortable in a wing chair and hastened to the immaculate kitchen where I arranged a cold compress with a thick towel. After taking off his shoe and sock, I found his ankle was considerably swollen, as I had suspected, and applied the compress despite his protests.

He was extremely grateful and, to my embarrassment, invited me to lunch.

The servant had returned, and I was slightly surprised to see a light-skinned Arab girl with doe eyes and a downcast manner. She was obviously upset to see her master apparently in pain and being attended by a stranger, but I soon put her at her ease. She was about twenty-five years of age, and he later told me she had been with him since the age of fifteen. He also said that he had settled a large sum of money on her in his will, in case anything happened to him, as he had no living relatives. I often wondered what the relationship was between him and the girl, but it was no business of mine.

But she prepared an excellent lunch which we washed down with an equally excellent Beaune, and I was amused to hear the Professor say that religious beliefs should never interfere with living standards.

Afterward the girl, whose name I never learned, produced an ancient wheelchair from an outhouse in the garden and I was able to wheel the Professor through his domain as he was anxious to show me around. Like all such houses which look like two storeys from outside, it was all on one level, due to the very high ceilings for coolness in that very hot climate. I was amazed at the size of his library, which contained about fifteen thousand books in many different languages. He knew the Koran inside-out, and had written many learned treatises on its aspects and teachings, expressing regret that its wisdom concerning forgiveness and tolerance had been distorted by some factions. All in all, the day had been a wonderful intellectual treat, and he expressed great regret when it came time to part. But dusk comes early in those parts, and the shadows were long on the ground when. I left, ushered to the door by the little dark-haired girl.

Mahmoud had asked for my telephone number and had said that I must come again soon. I had already warmed to his personality and had said that I would be delighted. As I went down the road, lined with its trim white villas and old gentlemen in white towelling robes on their way to the beach, I had much to think of.

The tottery figures which dotted the dusty roadway each had their own private beaches with elegant white pavilions which acted as changing rooms and where tea-things were kept for the frequent picnics which obtained there. I wondered idly how much longer this sybaritic existence would continue, in the constantly changing and unstable world which was now unfolding.

On my way to the tram depot I met an Egyptian police officer in his smart uniform and *képi* cap with its white *couvre nuque* to protect the back of the neck from the burning sun. I knew the police assigned officers to keep an eye on the properties in such rich neighbourhoods. The officer had a broad smile on his face as he gave me a salute.

"*Saida, effendi!*"

"*Saida, Monsieur l'agent!*" I said in return as I knew this particular officer spoke fluent French as well as Arabic.

I had a lot to think about as I jolted my way back in an ancient tram toward the heart of Alex.

II

It was a fortnight before I heard from the professor again. I had a telephone call one morning and he asked me to come to lunch the following day. He was quite recovered from the injury to his ankle and immensely grateful for my help.

Of course I accented and looked forward once again to being in the presence of so remarkable a man. I say remarkable advisedly because, although I had been in Mahmoud's company for only a few hours, I had realised he was incredibly learned not only in classical disciplines but in strange and esoteric byways of scholarship, and I was only too glad to be in his presence.

In the meantime I had been studying various reference works at the university and had discovered a number of remarkable facts illustrating the many facets of this extraordinary man's personality. Strangely, this documentation made no mention of the professor's forename; he had never volunteered it and I had never liked to ask as it might have offended such a courteous man. Perhaps he was merely looking for a friend on the same in intellectual level—though much diminished in my case—but nothing further.

In the meantime I had wandered round many of the tiny alleys that bisected hundreds of sections of this fascinating city; places where Europeans seldom penetrated but which were quite safe, as most Egyptians at that time were decent and law-abiding people. The *souks* and the streets full of coffee-houses were particularly interesting and I was beginning to familiarise myself with some of these areas, which wore absolutely baffling to the newcomer. Like the Professor, I had familiarised myself with these coffee-houses which followed strict hygienic practices, as one has to be careful in these matters.

Like many another before me, I was fascinated by the beauty and sweep of the great harbour and often frequented the native quarter where craftsmen built the *caiques*, the beautifully constructed fishing vessels that had plied these waters since time immemorial. I was also intrigued by the stalls where equally skilled shoemakers would create a pair of sandals which fitted perfectly from the raw leather in a matter of minutes, the stitching absolutely professional as though created by a machine. Beer Street was full of brothels with long queues of native customers waiting to enter. It was safe enough to walk through by day, on the way up from the docks, but was best avoided at night, where knife fights were frequent.

It rained only once in the three magical years I spent in the city and the days passed in a dream-like haze, so that I was glad that the advent of Professor Mahmoud would provide a more positive aspect in those hours spent away from my professional duties. Though I must not say that these apparently idle hours were entirely wasted. The big sailing vessels, with their lateen sails, manned by highly professional sailors, which traversed the Nile, and the many curious crafts that flourished in small premises in the *souks*, provided much material for my articles and essays. But I once had to warn two unsuspecting tourists who were trying to cross a rickety wooden bridge that spanned the unlikely named Sweetwater that ran through the city, that it was likely to collapse at any moment and if they fell into the water which was teeming with bacteria and other organisms that would result in a number of vile diseases, they would need expert medical advice and a number of painful injections.

In a city teeming with crippled beggars and many other poverty-stricken people who existed side-by-side with the most incredibly wealthy high class Egyptians and Levantines, the temptations were obvious. But I digress. Though I must add that my pair of naïïve tourists had been given an address where they could find a cheap hotel. It was a dragoman who had given them the information, but was obviously a scoundrel of the first water for the "hotel" was a notorious den in Sisters Street, where all the major brothels were situated. They were pathetically grateful and I was able to direct them to an excellent establishment on one of the main streets, where the proprietor was well known to me.

I never tired of watching the sun rise over the Ras el-Tin Palace, once the abode of the deposed Farouk, its sugar icing façççade, washed in warm light and reflected in undulating patterns in the harbour water. And I was always amused at the sight of the Egyptian training ship *Faroukieh*, modelled in steel on the lines of a vessel of Nelson's day, where youngsters were trained for the Egyptian Navy, which consisted of a few motor boats which patrolled the Great Bitter Lakes.

But these things were not the real Egypt so far as Mahmoud was concerned, merely tourist baubles to swell the coffers of the nation's exchequer. He was right, of course, but the romantic in me responded to such things which remained symbols of the country which millions around the world would never be able to visit. The reality behind the façççade was entirely different but no less valid. I thought I knew a great deal about Pompey's Pillar, one of the sights of the city, but Mahmoud explained a great deal more of its history than I would ever learn from a guide book.

Although I had been in the city for a year at this time, I had never visited the catacombs, a great underground city of the dead which was of Roman origin and lay deep below the throbbing heart of the modern city. The skeleton remains of the thousands of human beings which had once thronged the streets of the upper world were a grim reminder of mortality. It would have been unbearably sombre if alone, yet this great ossuary was not a morbid place, despite the stone shelves sullenly laden with the countless skeletons of those who had passed on centuries ago.

The rushing water of an underground river ran like a miniature Styx beneath an ancient stone bridge. The muted chatter of tourists, who passed like shadows beneath bare electric light bulbs, underlaid the atmosphere and did not interrupt the smooth flow of the Professor's learned discourse, enunciated in precise English, as he expounded on many things which the average guide would never have known.

With him I also went to the races, a new experience for me, with the Professor putting down large sums to back his favourite horses. It was a fine course, frequented by well-to-do people, with excellent spectator stands quite the equal of European venues. Racing was all Greek to me, as one might say. I knew nothing of form or which mounts were the favourites, but the Professor was in his element and seemed to have won a fair amount of money, and he insisted on buying me an excellent lunch afterward.

With him I ventured on a Felucca trip down the Nile, marvelling at the skills of the native helmsman and the crew, and on to the splendours of Karnak. And he took me on a guided tour of the great Cairo Museum, which was revealed in an extraordinary different light in view of the Professor's great knowledge. I knew I was receiving an education which I had never suspected, merely skimming the surface of Egyptian culture.

But one evening I jolted my way in one of the neolithic trams up to Sidi Bishr. I had an hour clear before I was due at Mahmoud's house for dinner, so I had resolved to explore the area a little further. As we passed station after station, I thought of the squalid association between the English novelist E.M. Forster and an Egyptian tram conductor on this very line.

After Sidi Bishr and the tramway turning area, the line petered out in the eternal sand.

As I went on it was extremely quiet except for the faint susurration of the sea in the distance and there was hardly anyone about. I walked on, my shadow dark-etched on the ground. I saw on a dune faraway the head of a camel coming into sight followed by its rider, who looked like a bearded Taureg with what looked like an antiquated rifle slung across the pommel of his saddle, and realised once again how thin the veneer of civilisation as understood in the West was upon this land once outside the confines of the cities.

I had once travelled to Cairo in what had been described as a luxury train but had found that the "toilet" consisted of two wooden cleats for the feet and a hole cut in the floor of the compartment.

Once again I found myself closer to the sea, and a number of Arabs had suddenly appeared as though from nowhere. There was nothing sinister in this, and I then perceived a white tower slowly emerged from the sand as I drew closer. A muezzin then began calling from the Minarat, "Allah, Akbar," literally the voice of God, and I realised it was time for evening prayers. Most of the Egyptians are deeply religious people and, unlike many of my compatriots, I respected their customs.

The white-robed Arabs around me had produced their prayer mats and were kneeling, facing Mecca. I stopped to attention and faced in the same direction until the prayers were over. I had seen ignorant Westerners laugh and walk on during such occasions and I deplored their appalling ignorance. I had seen some of them attacked for their loutish behaviour, and I really could not blame the reaction their ignorance had caused.

But it was now getting toward dusk, which comes early in those parts, and I retraced my steps until I was in sight of my host's villa, in anticipation of a stimulating evening. I was not disappointed.

After the excellent food and wine, and some three hours of Mahmoud's brilliant conversation, I once again felt I had been in the presence of a great man and one who was kind and generous, which is not always the case with such characters. As I jolted back toward the centre of the city on a late-night tram, I realised I had a great affection for these ancient vehicles, which made it so easy to get about.

I had some hilarious experiences during such trips. Those Arabs too poor to pay the fares would hang on to the sides and sometimes clamber over the roofs in order to get a free ride. The tram drivers and conductors tolerated this not only due to their sympathy for the poor but also because, due to the numbers of such free riders, they could do nothing about it.

But the extra passengers sometimes proved an advantage to the tram operators. Sometimes, at a crossing, trams would jump the lines. A great cheer would go up and all the Arabs would jump off and literally lift the vehicle back on to its tracks. They had a sense of humour rather like the British and would scream with laughter.

These street comedies were quite frequent and provided light relief both for the passengers and passers-by in the busy streets. There was never any mention of such things in sedate travel books.

Mahmoud's great erudition and his brief lectures on various aspects of Egyptian daily life were a constant source of enjoyment to me, and I only wish I could remember them in detail at this distance in time. My association with him was a form of education for a young man searching for knowledge and I shall always be grateful for his kindness and friendship over a long period of time. He always had the *Egyptian Mail*, the English newspaper, together with the two major Arabic newspapers, *Akbar el Yom* (literally "The Voice of the People") and *El Ahrarn* ("The Nation"), and his wise and pertinent comments on some of the stories in the three journals were an enriching experience for me.

He would laughingly proceed to pick to shreds the different interpretations the three newspapers drew from the same events. And he was always right. His own solutions were always pithy and to the point, pared to discard the superfluous verbiage and vacuous conclusions drawn by ignorant or bigoted correspondents regarding the events portrayed.

The Professor often told me of his doubts about the holy books of this world, sentiments that would have amounted to blasphemy in his own country. Perhaps he had become disillusioned in his later years, because he often mentioned that all the holy books of the world had been written by human beings, and that there had never been any proof of a Supreme Being, of whatever religion, and there was certainly no benevolence stemming from that, given the murderous chaos of the modern world with all its wars and disease. Despite this, he often amazed me by quoting enormously long passages from the Bible and (in translation) the Koran verbatim and without pause.

Perhaps his vast erudition had finally disturbed and even unhinged his magnificent brain, and that he longed for death. Who knows? Fate is an inscrutable master of men's destinies.

III

Mahmoud was a man of astonishing contrasts. I had imagined him to be a savant who was interested only in dead languages and ancient history, but he was astonishingly catholic in his tastes. I was amazed to find that he way a great enthusiast for the films of Laurel and Hardy, Buster Keaton and Harold Lloyd—he belonged to a society where such films were regularly shown—and the even wider range of his interests was extraordinary. One day he proposed that we should visit the M.G.M. cinema in Alexandria, which was a magnificent building and a showplace for the company's films It was in a central location, of course. As might be imagined, it was the most opulent of all the cinemas in the city and there were a great many them.

As usual, there was an immense foyer, like that of a live theatre, and a huge zinc bar with many waiters in immaculate white uniforms both serving behind the bar and bringing drinks to the wicker tables set about the gleaming mosaic floor. When one ordered wine or beer there was a sort of free lunch counter where one could eat salted biscuits, small cucumbers in vinegar, sandwiches, sardines, radishes and many other delicacies, supplied free when one ordered drinks.

There were separate performances, as in live theatres, and a bell rang to let waiting patrons know that the auditoria were emptying. The main auditorium was downstairs, I thought, but I was wrong. To my surprise we ascended by lift to the roof garden, which was open to the sky and one could see the lights of Alexandria with its lines of crawling traffic below. There was trelliswork with troughs of tropical flowers all round, which gave off a sensuous perfume. There was an enormous screen at the far end, and the raked seating gave one a perfect view. As normal, all the seating was of wickerwork, as otherwise extremely dangerous insects would have inhabited the upholstery. The films were shown far into the night, as the city never seemed to sleep. As in all the cinemas in Alexandria, the films were shown in English, with French sub-titles while a large box-like screen to the left of the main screen displayed the dialogue in Arabic. Extremely ingenious, I thought, and the Professor who was, of course, proficient in all three languages, delighted in the different interpretations of the actual dialogue, which were displayed.

When we came out with the crowd and went downstairs, I was amused to see how M.G.M. had developed a simple but profitable system. The programmes in the upstairs and downstairs auditoria were staggered by about twenty-five minutes. As soon as the first reels of a feature were screened in the downstairs salle, they were speedily rewound by machine and a small boy, staggering with the weight of the cans, took them up in the lift to the roof garden projection box for screening. The system worked perfectly, and I was always intrigued to see how smoothly this system worked on my numerous subsequent visits.

But I little suspected then that the golden times in the Professor's company were about to come to an abrupt end. I did not see the Professor for a while after the last of our series of visits to the cinema, as I was extremely busy and so was he, but we often spoke on the telephone. It was about a month after our last meeting that I received another invitation to lunch. But in fact the friendship was about to end in an abrupt way which no-one could have expected.

*

IV

I travelled out on a ramshackle tram along the familiar line to Sidi Bishr. I had bought him a special bottle of cognac which was a special pleasure to Mahmoud, who always drank a small glass with strong coffee after the two main meals of the day.

I sensed there was something wrong when I arrived at my destination. Several vehicles were parked some way away along the dusty street, which was unusual as there was no house at that location. When I rang the bell, a gigantic figure opened the door. I recognised the Chief of Police and, as he had his fez on, I knew he was on official business. I knew him well, of course, as I had written a number of articles on the work of the bureau.

"Something is wrong?"

He nodded grimly. "Very serious I'm afraid." Like Mahmoud, he spoke excellent English. "You'd better come in, *effendi*."

I followed him into the reception hall where two plainclothes detectives were sitting at a table studying documents. Just then the telephone rang and the nearest of the men picked it up and started speaking in Arabic.

"We'd better go into the salon," my companion said.

"What on earth has happened?" I asked as we seated ourselves on two large carved wooden chairs.

The officer shrugged. "The Professor is dead, I regret to say."

I was stunned. "What...? How...?" I stammered. "He was fine the last time I saw him. A heart attack...?"

The big police officer shook his head. "His throat has been cut."

I looked at him in shocked disbelief. "Suicide then?"

Again the vehement shake of the head. "Murder, *effendi*. From the position of the wound, our doctor says he was attacked from behind."

I could not speak, I was so overcome. It was the last thing I had expected.

"You had better drink some of that cognac you are carrying. I will fetch a glass."

He brought back a crystal goblet from a cabinet and I unsealed the bottle and poured myself a stiff one. I knew this officer did not drink alcohol.

"Do you know who is responsible?"

Again the grave inclination of the head. "The gardener. He will not get far."

I could not really take all this in. "And if you catch him, what then?"

"Death," he said grimly. "The professor was a very distinguished man and one of the city's most celebrated citizens."

"I still do not understand..."

The big officer laid his hand on my arm. "The murderer is the girl's husband."

I could not take this in for a few seconds. Then I was shattered and poured myself another stiff peg. "But how…?"

My informant spread his hands wide. "It is perfectly simple, *effendi*. The girl had been the Professor's mistress. She got this man his job as a gardener. Jealousy eventually overcame his patience."

He held up his hand to forestall my next question. "The Professor had made this house and all his wealth over to the girl as his sole beneficiary. For all her years of loyal service. Bizarre, is it not?"

I did not reply. I was still numbed with shock.

"We have cross-examined the girl for hours," the officer went on, "and we are now convinced that she knew nothing of the inheritance or what her husband intended to do. She is still in shock, and a nurse from the local hospital will be here shortly.

"We discovered the Professor's will, properly attested, among his papers. Though the girl certainly did not know about this, it seems to us that the Professor may have innocently informed the gardener of his intentions, with the terrible result we have seen.

"This man was utterly stupid," continued the officer. "He would have shared with his wife if he had waited. All that knowledge and that great brain stilled forever. Helas…"

I got up to go, still staggered, and he took my arm and led me to the door.

"Despite his great learning, in some ways the Professor was an incredibly innocent man."

I could not but concur with the officer's remarks. I thanked him and left.

All the way back to Ramleh Square in the wobbly tram I could not help thinking of all the good times we had shared together and all the distilled wisdom of that great mind.

V

The *Egyptian Mail* had a field day, of course. I was too sick at heart to read all the details. I had lost the friendship of the wisest man I had ever known.

Ten days later I attended his simple funeral. He was buried in the European section of the cemetery, as he had stipulated in his will. He was only sixty years old, but I had felt him to be much older.

He had left a letter for me, written several weeks before his death, in which he expressed his delight at all the good times we had had together. I finally realised that he had been a very lonely individual.

The funeral was simple, with an English padre conducting the service. There were all of thirty people there; several elderly colleagues from the various faculties in which he had conducted seminars and lectures. No European women. I sensed that many people were disgusted at the lurid stories that had been circulating in the press. But then, in their enclosed circles, they knew little about human nature.

I was more tolerant, and so were some of the younger colleagues of the Professor, who had gathered to pay their tributes in memory of a great man.

The girl was there too, standing aloof and apart at the back of the gathering. When I glanced at her, she avoided my eyes, so I did not approach her as we dispersed. Her husband was in prison under sentence of death. Justice is swift in those parts.

VI

I settled up all my affairs and left Alexandria three months later. Nothing could ever be the same.

Ras el-Tin Palace was like a mirage or a dream, as the late afternoon sun gilded its extravagant outlines. A man in dark clothing was walking along the harbour wall, and he raised his arm in salute as the ship headed out to sea. I like to think it was the Professor's final benediction.

The Chief of Police had given me a photograph of him; a fine portrait study which was dedicated to me in the warmest terms, in his immaculate handwriting. Had he known what was to happen and had wanted to give me a memento before it was too late? He was playing a dangerous game which ended in disaster. Either way, I shall never know. But I keep it in pride of place at our house in Kent, and my wife and I often discuss him.

Since then I have never ceased to think of the Professor and all that world of knowledge so quickly snuffed out. But the friendship he had extended to me and his wise words still reverberate in my ears. I dismiss all the other things I have heard about him. For my part, I have never had a truer friend, and the great wisdom he has bestowed on me with the richness of his mind still leaves a fragrant memory.

Ras el-Tin slowly dissolved in the fading dusk as the amazing country of Egypt disappeared from sight and the liner headed out to sea, making course for England through the velvet night. That moment of farewell held a fragrance that is with me still.

G. SHORT STORIES & NOVELLAS

Includes an alphabetical listing of short stories and novellas written by Basil Copper and published in magazines and books

(*"The procedure when writing fantasy, horror or a macabre novel or short story is entirely different. These must be plotted meticulously and I always begin with the finale first. Obviously, I don't write it first, but put it down on a piece of paper together with any other colour notes that may be applicable. Then, when one sits down to write the short story or novella, one is always working towards that finale, which should convey great shock and surprise. I try and eschew the more crude and horrific effects that so many of the modern trash merchants go in for; a verbal equivalent of Hammer Films at their goriest. In my books, less is more, which is why I admire those writers of the past like E.F. Benson, M.R. James, H.P. Lovecraft, whose subtle terrors were so far more chilling than most of today's so-called horror merchants could achieve. The footfall late at night, fingers at the window, the laugh at the top of the stairs."* BC.)

G1 THE ACADEMY OF PAIN

a. __________, in *Legends for the Dark: Tales of Fantasy and Horror* edited by Peter Haining, New English Library, February 1968 (pb).

b. [as **LA ACADEMIA DEL DOLOR**], in *Antologíia de relatos de espanto y terror 5*, Ediciones Dronte, Spain, 1972 (tp)*.

c. __________, in *Summoned from the Tomb: Great Tales of Horror* edited by Peter Haining, Sidgwick & Jackson, London, England, 1973 (hc).

[Note: 'The Academy of Pain' does not appear in the 1966 edition of this book, which has different contents.]

d. __________, see When Footsteps Echo: Tales of Terror and the Unknown [B7].

e. __________, in *Dark Voices 3: The Pan Book of Horror* edited by David Sutton and Stephen Jones, Pan Books, London, England, 1991 (pb).

G2 THE ADVENTURE OF THE ANGUISHED ACTOR

a. __________, see Solar Pons #10: The Secret Files of Solar Pons [C7].

b. __________, see Solar Pons The Final Cases [C9].

G3 THE ADVENTURE OF THE BAFFLED BARON

a. __________, see Solar Pons The Final Cases [C9].

G4 THE ADVENTURE OF BUFFINGTON OLD GRANGE

a. __________, see Solar Pons #8: The Dossier of Solar Pons [C2a].

b. __________, see The Dossier of Solar Pons [C2b].

G5 THE ADVENTURE OF THE CALLOUS COLONEL

a. __________, see The Exploits of Solar Pons [C3].

b. __________, in *Murder Most Scottish* selected by Stefan Dziemianowicz, Bab Adey, Ed Gorman and Martin H. Greenberg, Barnes & Noble Books, New York, New York, 1999 (hc).

G6 THE ADVENTURE OF THE CRAWLING HORROR

a. __________, see Solar Pons #10: The Secret Files of Solar Pons [C7].

b. __________, in *Dark Detectives: Adventures of the Supernatural Sleuths* edited by Stephen Jones, F&B Mystery, Minneapolis, Minnesota, 1999 (hc).

There are two varients of this edition.

i. Description: 2,000-copy edition.

ii. Description: 100-copy signed and slipcased edition.

G7 THE ADVENTURE OF THE CURSED CURATOR

a. __________, see The Recollections of Solar Pons [C6].

G8 THE ADVENTURE OF THE DEFEATED DOCTOR

a. __________, see Solar Pons #9: The Further Adventures of Solar Pons [C4].

b. __________, see The Further Adventures of Solar Pons [C5].

G9 THE ADVENTURE OF THE FRIGHTENED GOVERNESS

a. __________, see Solar Pons #11: The Uncollected Cases of Solar Pons [C11].

b. __________, see The Further Adventures of Solar Pons [C5].
c. __________, in *The Mammoth Book of Historical Detectives* edited by Mike Ashley, Robinson Publishing, London, England, 1995 (tp).
d. __________, in *The Mammoth Book of Historical Detectives* edited by Mike Ashley, Carroll & Graf Publishers, Inc., New York, New York, 1995 (tp)
e. [as **DAS ABENTEUER DER VERÄNGSTIGTEN GOUVERNANTE**] in *Räuber Schurken Lumpenpack: Weitere Historische Kriminalgeschichten* edited by Mike Ashley, Bastei Lübbe, Berlin, Germany, April 1998 (tp)*.
f. __________, in *Historical Detectives* edited by Mike Ashley, Castle Books, Edison, New Jersey, 2002 (hc)*.
g. [as **DAS ABENTEUER DER VERÄNGSTIGTEN GOUVERNANTE**] in *Räuber Schurken Lumpenpack: Historische Kriminalgeschichten* edited by Mike Ashley, Alle Recht, Berlin, Germany, May 2004 (hc)*.

G10 **THE ADVENTURE OF THE HAMMER OF HATE**
a. __________, see Solar Pons #8: The Dossier of Solar Pons [C2a].
b. __________, see The Dossier of Solar Pons [C2b].

G11 **THE ADVENTURE OF THE HAUNTED RECTORY**
a. __________, see Solar Pons #11: The Uncollected Cases of Solar Pons [C11].
b. __________, see Solar Pons The Final Cases [C9].

G12 **THE ADVENTURE OF THE HOUND OF HELL**
a. __________, see The Recollections of Solar Pons [C6].

G13 **THE ADVENTURE OF THE HORRIFIED HEIRESS**
a. __________, see Solar Pons #10: The Secret Files of Solar Pons [C7].
b. __________, see Solar Pons The Final Cases [C9].

G14 **THE ADVENTURE OF THE IGNORED IDOLS**
a. __________, see Solar Pons #10: The Secret Files of Solar Pons [C7].
b. __________, see Solar Pons The Final Cases [C9].

G15 **THE ADVENTURE OF THE IPI IDOL**
a. __________, see Solar Pons #8: The Dossier of Solar Pons [C2a].
b. __________, see The Dossier of Solar Pons [C2b].

G16 THE ADVENTURE OF THE MAD MILLIONAIRE

a. __________, see The Recollections of Solar Pons [C6].

G17 THE ADVENTURE OF THE MISSING STUDENT

a. __________, see Solar Pons #9: The Further Adventures of Solar Pons [C4].

G18 THE ADVENTURE OF THE PERPLEXED PHOTOGRAPHER

a. __________, see Solar Pons #8: The Dossier of Solar Pons [C2a].

b. __________, see The Dossier of Solar Pons [C2b].

G19 THE ADVENTURE OF THE PERSECUTED PAINTER

a. __________, in *The Mammoth Book of New Sherlock Holmes Adventures* edited by Mike Ashley, Robinson, London, England, 1997 (tp).

b. __________, in *The Mammoth Book of New Sherlock Holmes Adventures* edited by Mike Ashley, Carroll & Graf, New York, New York, 1997 (tp).

c. [as **DER GEPLAGTE MALER**], in *Sherlock Holmes und der Fluch von Addleton: Neue Sherlock-Holmes-Geschichten* edited by Mike Ashley, Bastei Lübbe/Verlagsgruppe Lübbe, Bergisch Gladbach, Germany, June 2003 (tp).

d. __________, in *New Sherlock Holmes Adventures*, edited by Mike Ashley, Castle Books, Edison, New Jersey, 2004 (hc)*.

e. __________, see Solar Pons The Final Cases [C9].

G20 THE ADVENTURE OF THE PHANTOM FACE

a. __________, see The Exploits of Solar Pons [C3].

G21 THE ADVENTURE OF THE SHAFT OF DEATH

a. __________, see Solar Pons #9: The Further Adventures of Solar Pons [C4].

b. __________, in The Further Adventures of Solar Pons (1987).

G22 THE ADVENTURE OF THE SINGULAR SANDWICH

a. __________, see Solar Pons #11: The Uncollected Cases of Solar Pons [C11].

b. __________, see The Adventure of the Singular Sandwich [C1].

c. __________, see The Recollections of Solar Pons [C6].

G23 **THE ADVENTURE OF THE SIX GOLD DOUBLOONS**

a. __________, see Solar Pons #8: The Dossier of Solar Pons [C2a].

b. __________, see The Dossier of Solar Pons [C2b].

G24 **THE ADVENTURE OF THE SURREY SADIST**

a. __________, see Solar Pons #9: The Further Adventures of Solar Pons [C4].

G25 **THE ADVENTURE OF THE VERGER'S THUMB**

a. __________, see The Exploits of Solar Pons [C3].

G26 **AMBER PRINT**

a. __________, in *Dr. Caligari's Black Book* edited by Peter Haining, New English Library, London, England, November 1969 (pb).
[Note: 'Amber Print' does not appear in the 1968 edition of this book, which has different contents.]

b. __________, see From Evil's Pillow [B3].

c. __________, see When Footsteps Echo: Tales of Terror and the Unknown [B7].

d. __________, in *Christopher Lee's 'X' Certificate* edited by Christopher Lee and Michel Parry, W.H. Allen, London, England, 1976 (hc).

e. __________, in *Christopher Lee's 'X' Certificate* edited by Christopher Lee and Michel Parry, Star Books/W.H. Allen, London, England, 1975 (pb).

f. __________, in *From the Archives of Evil* edited by Christopher Lee and Michel Parry, Warner Books, New York, New York, 1976 (pb).

g. __________, in *The Mammoth Book of Terror* edited by Stephen Jones, Carroll & Graf, New York, New York, 1991 (tp).

h. __________, in *The Mammoth Book of Terror* edited by Stephen Jones, Robinson Publishing, London England, 1991 (tp).

i. __________, in *The Anthology of Horror Stories* edited by Stephen Jones, Tiger Books International, Twickenham, England, 1994 (hc).

j. __________, in *The Giant Book of Horror* edited by Stephen Jones, The Book Company, Sydney, Australia, 1996 (tp).

k. __________, in *The Giant Book of Horror* edited by Stephen Jones, Parragon, Avonmouth, Bristol, 1996 (tp).

l. __________, in *The Giant Book of Horror* edited by Stephen Jones, Magpie Books, London, England, 1996 (tp).

m. [as **L'ESEMPLARE COLOR SEPPIA**], in *Terrore!* edited by Stephen Jones, Newton & Compton, Rome, Italy, 1996 (tp).

G27 **ARCHIVES OF THE DEAD**

a. __________, in *The Evil People; Being Thirteen Strange and Terrible Accounts of Witchcraft, Black Magic and Voodoo* edited by Peter Haining, Leslie Frewin Publishers Limited, London, England, 1968 (hc).
b. __________, in *The Evil People; Being Thirteen Strange and Terrible Accounts of Witchcraft, Black Magic and Voodoo* edited by Peter Haining, Popular Library, New York, New York, 1968 (pb).
c. __________, see When Footsteps Echo: Tales of Terror and the Unknown [B7].
d. __________, in *The Evil People* edited by Peter Haining, Everest Books Limited, London, England 1975 (pb).
e. __________, see And Afterward the Dark: Seven Tales [B1].

G28 **AS THE CROW FLIES**

a. __________, see Cold Hand on My Shoulder: Tales of Terror & Suspense [B2].

G29 **BETTER DEAD**

a. __________, in *The Mammoth Book of Frankenstein* edited by Stephen Jones, Robinson Publishing, London, England, 1994 (tp).
b. __________, in *The Mammoth Book of Frankenstein* edited by Stephen Jones, Carroll & Graf, New York, New York, 1994 (tp).
c. __________, in *The Mammoth Book of Frankenstein* edited by Stephen Jones, Justsystem, Japan, 1995 (hc).
d. __________, in *The Giant Book of Frankenstein* edited by Stephen Jones, Magpie Books, London, England, 1995 (tp).
e. __________, in *The Giant Book of Frankenstein* edited by Stephen Jones, The Book Company, Sydney, Australia, 1995 (tp).
f. [as **MEGLIO MORTO**], in *Tutte le Storie de Frankenstein* edited by Stephen Jones, Newton Compton, Rome, Italy, 1996 (tp).
g. __________, see Whispers in the Night: Stories of the Mysterious and Macabre [B9].

G30 **BEYOND THE REEF**

a. __________, in *Shadows Over Innsmouth* edited by Stephen Jones,

Fedogan & Bremer, Minneapolis, Minnesota, 1994 (hc).

There are two varients of this edition.

i. Description: 2,000-copy edition.

ii. Decription: 100-copy signed and slipcased edition.

b. __________, in *Shadows Over Innsmouth* edited by Stephen Jones, Victor Gollancz, London, England, 1997 (tp).

c. __________, in *Shadows Over Innsmouth* edited by Stephen Jones, Del Rey/Ballantine Books, New York, New York, September 2001 (tp).

d. __________, in *Shadows Over Innsmouth (Volume 1)* edited by Stephen Jones, Gakken, Japan, 2001 (pb).

e. [as **DAS GEHEIMNIS VON INNSMOUTH**], Festa, Almersbach, Germany 2002 (hc). [Note: #1 in "H.P. Lovecrafts Bibliothek des Schreckens" series. Published in a 400-copy signed edition.]

f. __________, see Cold Hand on My Shoulder: Tales of Terror & Suspense [B2].

(*"I wanted the atmosphere to be of a 'faded 1920s variety' as in HPL's original, but without being too much of a pastiche or derivative in any way."* BC.)

G31 **THE BLACK COFFIN**

a. __________, see this volume.

G32 **BRIGHT BLADES GLEAMING**

a. __________, in *Dark Love* edited by Nancy A. Collins, Edward E. Kramer and Martin H. Greenberg, Penguin Books/ Roc, New York, New York, November 1995 (hc).

b. __________, in *Dark Love* edited by Nancy A. Collins, Edward E. Kramer and Martin H. Greenberg, Penguin Books/ Roc, New York, New York, August 1996 (pb).

c. __________, in *Dark Love* edited by Nancy A. Collins, Edward E. Kramer and Martin H. Greenberg, Hodder & Stoughton, London, England, 1995 (tp).

d. __________, in *Dark Love* edited by Nancy A. Collins, Edward E. Kramer and Martin H. Greenberg, New English Library/Hodder & Stoughton, London, England, 1995 (pb).

e. __________, in *Dark Love: Twenty Two All-Original Tales of Love and Obsession* edited by Nancy A. Collins, Edward E. Kramer and Martin H. Greenberg, FusoSha Mystery, Japan, May 1996*.

f. [as **CUCHILLAS RELUCIENTES**], in *Malignos y Macabros: 22 relatos para tener*

pesadillas edited by Edward E. Kramer, Nancy A. Collins and Martin H. Greenberg, Plaza y Janés/Jet, Spain 1997 (pb)*.
g. [as **LAME LUCENTI**], in *Dark Love*, edited by Nancy A. Collins, Edward E. Kramer and Martin H. Greenberg, Sperling & Kupfer Editori, Milan, Italy, February 1997 (hc)*.
h. [as **LA LUEUR ÉCLATANTE DES LAMES**], in *Noir comme l'amour* edited by Nancy A. Collins, Martin H. Greeneberg and Edward E. Kramer, le Grande livre du mois, Paris, France, January 1998 [December 1997] (hc)*.
i. [as **LA LUEUR ÉCLATANTE DES LAMES**], in *Noir comme l'amour* edited by Nancy A. Collins, Martin H. Greeneberg and Edward E. Kramer, Albin Michel, Paris, France, January 1998 [December 1997] (pb)*.
j. [as **LA LUEUR ÉCLATANTE DES LAMES**], in *Noir comme l'amour* edited by Nancy A. Collins, Martin H. Greeneberg and Edward E. Kramer, Librairie Générale Française/le Livre de poche, Paris, France, October 2000 (pb)*.
k. [as **LAME LUCENTI**], in *Dark Love* edited by Nancy A. Collins, Edward E. Kramer and Martin H. Greenberg, Sperling & Kupfer Editori, Milan, Italy, 2000 (pb)*.
l. _________, see Cold Hand on My Shoulder: Tales of Terror & Suspense [B2].

G33 **CAMERA OBSCURA**

a. _________, in *The Sixth Pan Book of Horror Stories* edited by Herbert Van Thal, Pan Books, London, England, 1965 (pb).
b. _________, in *Alfred Hitchcock Presents: Stories That Scared Even Me* [edited by Robert Arthur], Random House, New York, New York, 1967 (hc).
c. _________, in *Alfred Hitchcock Presents Stories That Scared Even Me* [edited by Robert Arthur], Max Reinhardt, London, England, 1967 (hc).
d. _________, see Not After Nightfall: Stories of the Strange and the Terrible [B6].
e. _________, in *Alfred Hitchcock Bringt: Mörderisches Bettgeflüster*, Wilhelm Heyne Verlag, München, West Germany, 1969 (pb).
f. _________, in *Alfred Hitchcock Presents Stories That Scared Even Me Part One*, Pan Books, England, 1970 (pb).
g. _________, in *Alfred Hitchcock Presents Scream Along With Me: Tales from Stories That Scared Even Me* edited by Alfred Hitchcock [and Robert Arthur], Dell Publishing Co., New York, New York, December 1970 (pb).
[NOTE: Listed as "Basil Cooper" on the back cover.]
h. [as **LA CÁMARA OSCURA**], *Relatos que me asustaron* edited by Alfred Hitchcock, Aguilar, Spain, 1972 (tp)*.

i. __________, see When Footsteps Echo: Tales of Terror and the Unknown [B7].
j. __________, see And Afterward the Dark: Seven Tales [B1].
k. __________, in *Hitchcock prééesente Histoires Terrifiantes*, Presses Pocket, Paris, France, February 1979 (pb).
l. [as **LA CÁMARA OSCURA**], *Relatos que me asustaron* edited by Alfred Hitchcock, Mundo Actual de Ediciones-Discolibro, Barcelona, Spain, 1980 (tp)*.
m. [as **LA CÁMARA OSCURA**], *Relatos que me asustaron* edited by Alfred Hitchcock, Ciírculo de Lectores/Sin Colección, Spain, 1988 (tp)*.
n. __________, in *Dark Voices: The Best from The Pan Book of Horror Stories* edited by Stephen Jones and Clarence Paget, Pan Books, London, England, 1990 (hc).
o. __________, in *Dark Voices: The Best from The Pan Book of Horror Stories* edited by Stephen Jones and Clarence Paget, Pan Books, London, England, 1990 (pb).

G34 **THE CANDLE IN THE SKULL**

a. __________, in *Hallowe'en Hauntings: Stories About the Most Ghostly Night of the Year* edited by Peter Haining, William Kimber, London, England, 1984 (hc).
b. __________, in *1993 World Fantasy Convention*, 1993 World Fantasy Convention, Minneapolis, Minnesota, October 1993 (convention souvenir book).
[Note: Basil Copper was Special Guest at this convention.]
c. __________, in *The Mammoth Book of New Terror* edited by Stephen Jones, Robinson, London, England, 2004 (tp).
d. __________, in *The Mammoth Book of New Terror* edited by Stephen Jones, Carroll & Graf, New York, New York, 2004 (tp).
e. __________, see Knife in the Back: Tales of Twilight and Torment [B5].

G35 **THE CAVE**

a. __________, see Not After Nightfall: Stories of the Strange and the Terrible [B6].
b. __________, see And Afterward the Dark: Seven Tales [B1].
c. __________, in *Summer Chills: Strangers in Stranger Lands* edited by Stephen Jones, Carroll & Graf, New York, New York, 2007 (tp).

G36 **CHARING CROSS-DOVER-CHARING CROSS**

a. __________, see Cold Hand on My Shoulder: Tales of Terror & Suspense [B2].

G37 **CHARON**

a. __________, see Not After Nightfall: Stories of the Strange and the Terrible [B6].

b. __________, see From Evil's Pillow [B3].
c. __________, see Voices of Doom: Tales of Terror and the Uncanny [B8].

G38 **CRY WOLF**

a. __________, see *Vampires, Werewolves and Other Monsters* edited by Roger Elwood, Curtis Books, New York, New York, 1974 (pb).
b. __________, see When Footsteps Echo: Tales of Terror and the Unknown [B7].
c. __________, in *Werewolf: Horror Stories of the Man-Beast* edited by Peter Haining, Severn House Publishers, London, England, 1987 (hc).
d. __________, in *The Mammoth Book of Werewolves* edited by Stephen Jones, Robinson Publishing, London, England, 1994 (tp).
e. __________, in *The Mammoth Book of Werewolves* edited by Stephen Jones, Carroll & Graf, New York, New York, 1994 (tp).
f. __________, in *The Giant Book of Werewolves* edited by Stephen Jones, Magpie Books, London, England, 1995 (tp).
g. __________, in *The Giant Book of Werewolves* edited by Stephen Jones, The Book Company, Sydney, Australia, 1995 (tp).
h. __________, in *The Giant Book of Werewolves* edited by Stephen Jones, Parragon, London, England, 1995 (tp).
i. [as **GRIDARE AL LUPO**], in *Lupi Mannari!* Edited by Stephen Jones, Newton & Compton, Rome, Italy, 1997 (tp).

(*"My only short story on the theme of lycanthropy, 'Cry Wolf', was published by Robert Hale in my collection* When Footsteps Echo *in 1975, while at the same time appearing under the aegis of St. Martin's Press in New York. For this wintry tale set amongst superstitious peasants in Eastern Europe, I did not reveal the identity of the werewolf until the end of the narrative, and the tale has received a number of airings in various anthologies since."* BC.)

G39 **THE CURSE**

a. __________, in *S~X~R: The Magazine of The Tonbridge Senior Boys School,* No.5, March 1938, Tonbridge, Kent, England (magazine).
[Note: As by "B. Copper". The same issue not only informs us that the young author was a member of The Leicester football eleven and that he received a "3rd Year Certificate" in 'Mr. Harvey's Temperance Lecture', but it also contains the following (unassigned) review of a Christmas play, produced by Mr. Graville: "Leicester House gave the most ambitious show in presenting a playlet, 'The Little Angel' (hero, Mr.

Stonely's little angel Basil Copper). D. Burr, with a wig nearly as big himself, made a fascinating blonde, worthy spouse of P. Johnson, and mother of the little angel, Willy. Boyce spoke out clearly as an explorer-uncle, spinning Baron Munchausen adventures, and Foster as the Prologue remembered his words more often than not. The play told the story of a bad little boy who ate some magic beans and was thereupon converted into a perfect 'little angel'. But the cure for which his parents had always longed, proved more distressing than the disease, and they were thankful when he returned to his former self.

'Little Willy' (to us Basil Copper),
Whose behaviour was, oh, most improper,
By the aid of a Bean
His wrong-doing has seen
And intends to be good – what a whopper."

b. __________, see this volume.

G40 **DEATH AT THE METROPOLE**

a. __________, see The Exploits of Solar Pons [C3].

G41 **DEATH OF A DEMI-GOD**

a. __________, in *Dark Destiny: Proprietors of Fate* edited by Edward E. Kramer, White Wolf Publishing, Clarkston, Georgia, 1995 (hc).

b. __________, in *Dark Destiny* edited by Edward E. Kramer, White Wolf Publishing, Clarkston, Georgia, 1995 (pb).

c. __________, see Cold Hand on My Shoulder: Tales of Terror & Suspense [B2].

("*'Death of a Demi-God' first appeared in the US in an edited and altered form in 1994. This is its first appearance with the definitive text.*" BC.)

G42 **DEATH OF A NOBODY**

a. __________, see Knife in the Back: Tales of Twilight and Torment [B5].

G43 **DIAMONDS ARE FOR NEVER**

a.__________, see this volume.

G44 **DOCTOR PORTHOS**

[Note: August Derleth contracted for this story in the early 1970s. It was intended to appear in *From Evil's Pillow*, and later scheduled for an edition of *The Arkham Collector*, but Derleth died before it could be published by Arkham House.]

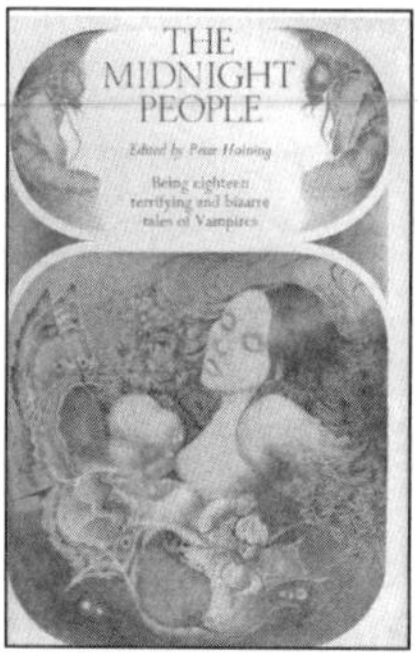

a. _________, in *The Midnight People: Being Eighteen Terrifying and Bizarre Tales of Vampires* edited by Peter Haining, Leslie Frewin Publishers Limited, London, England, 1968 (hc).
a. _________, in *The Midnight People: Being Eighteen Terrifying and Bizarre Tales of Vampires* edited by Peter Haining, Popular Library, New York, New York, 1968 (pb).
b. _________, in *Vampires at Midnight: Seventeen Brilliant and Chilling Tales of the Ghastly Bloodsucking Undead* edited by Peter Haining, Grosset & Dunlap, New York, New York, 1970 (hc). [Note: "Formerly titled *The Midnight People*".]
c. _________, in *The Midnight People* edited by Peter Haining, Everest Books, London, England, 1975 (pb).
d. _________, see When Footsteps Echo: Tales of Terror and the Unknown [B7].
e. [as **LE DOCTEUR PORTHOS**], in *Trois saigneurs de la nuit* edited by Jacques Finné, Nouvelles éditions Oswald, Paris, France, January 1986 (pb)*.
f. _________, in *The Mammoth Book of Vampires* edited by Stephen Jones, Robinson Publishing, London, England, 1992 (tp).
g. _________, in *The Mammoth Book of Vampires* edited by Stephen Jones, Carroll & Graf, New York, New York, 1992 (tp).
h. _________, in *Vampires at Midnight: Sixteen Brilliant and Chilling Tales of the Ghastly Bloodsucking Undead* edited by Peter Haining, Warner Books, London, England, April 1993 (pb).
i. _________, in *The Giant Book of Vampires* edited by Stephen Jones, Magpie Books, London, England, 1994 (tp).
j. _________, in *The Giant Book of Vampires* edited by Stephen Jones, The Book Company, Sydney, Australia, 1994 (tp).
k. _________, in *The Giant Book of Vampires* edited by Stephen Jones, Parragon, London, England, 1994 (tp).
l. _________, in *Book of Vampires* edited by Stephen Jones, Barnes & Noble Books, New York, New York, 1997 (hc).
m. [as **IL DOTTOR PORTHOS**], in *Vampir!* Edited by Stephen Jones, Newton & Compton, Rome, Italy, 1997 (tp).
n. _________, in *The Mammoth Book of Vampires* edited by Stephen Jones, Robinson, London, England, 2004 (tp). [NOTE: This is a revised version of the 1992 edition.]
o. _________, in *The Mammoth Book of Vampires* edited by Stephen Jones, Carroll

& Graf, New York, New York, 2004 (tp).
[NOTE: This is a revised version of the 1992 edition.]
p. __________, in *The Mammoth Book of Vampires* edited by Stephen Jones, Carroll & Graf, New York, New York, 2005 (hc).
[Note: Book Club edition. This follows the contents of the revised 2004 edition.]
q. __________, in *Vampires* edited by Stephen Jones, Azbooka-klassika, St. Petersburg, Russia, 2007 (hc).
[Note: This follows the contents of the revised 2004 edition of *The Mammoth Book of Vampires*.]

(*"I myself in my own writing career have employed restraint when using the vampire theme. I have written only one short story in the genre, 'Doctor Porthos', in which the narrator turned out to be the vampiric source of the horror. I felt that was the only way to squeeze a new thrill out of a very old situation. That I was successful was proved when Universal Pictures selected the story as one of my tales to be filmed in their famous* Night Gallery *series. The vagaries of the writer's life were nowhere more vividly illustrated when the story, together with three others from my pen, were passed for filming, only to be shelved by the front office when the script-writer and presenter Rod Serling died following a massive heart attack."* BC.)

G45 DUST TO DUST

[Note: This was originally intended to be published in booklet form.]
a. __________, in *Midnight Sun*, Number 4, August 1976, edited by Gary Hoppenstand, USA (magazine).
b. __________, see And Afterward the Dark: Seven Tales [B1].

G46 LES ÉTRANGES SŒURS DUBOIS

a. __________, in *Confidences* No.1280, May 23 1972, Paris, France (magazine).
[Note: Extract from a "Mike Faraday" novel.]

G47 FINAL DESTINATION

a. __________, see Whispers in the Night: Stories of the Mysterious and Macabre [B9].

G48 THE FLABBY MEN

a. __________, see And Afterward the Dark: Seven Tales [B1].
b. __________, in *The Urbanite* No.9: *Strange Places*, Fall 1997, edited by Mark McLaughlin, Urban Legend Press, Davenport, USA (magazine).
[Note: "Featured Writer: Basil Copper".]
c. __________, in *The Mammoth Book of Monsters* edited by Stephen Jones, Robinson Publishing, London, England, 2007 (tp).
d. __________, in *The Mammoth Book of Monsters* edited by Stephen Jones, Robinson Publishing, London, England, 2007 (tp).

G49 THE GOSSIPS

a. __________, see From Evil's Pillow [B3].
b. __________, see Voices of Doom: Tales of Terror and the Uncanny [B8].

c. __________, in *Keep Out the Night* edited by Stephen Jones, PS Publishing, Harrogate, England, October 2002 (hc).
There are two varients of this edition.
i. Description: 500-copy edition signed by the editor.
ii. Description: 100-copy slipcased deluxe editions signed by all the contributors.

(*"Like most authors, one's ideas seem to come naturally, by a strange process–rather like osmosis–whereby fragmentary images surface in one's consciousness. Many years ago, during World War II, when I was on the communications staff on a huge depot ship in Alexandria harbour, I went ashore one day and met a chap I knew in the bar of the Fleet Club. He was a fellow radio operator and had been stationed on a wireless station in Sicily, after the war had passed on. He mentioned that it was an extremely strange place away from the towns–baking heat; oppressive silences; hidden villages where no-one seemed to stir; half-ruined palazzos in certain areas; the only sound the buzzing and chirping of insects where there was grass and foliage.*

"This, combined with the remembrance of Canova's statue 'The Three Graces' came back into my mind when I was compiling a collection of macabre tales years ago which I intended to call The Gossips and Other Queer Tales. *I thought it might be interesting if the narrator heard whispering going on, as though emanating from the statues, and things progressed from there.*

"Such stuff as dreams are made of, of course, but none the worse for that." BC.)

G50 THE GRASS

a. _________, see Whispers in the Night: Stories of the Mysterious and Macabre [B9].

[Note: Basil Copper's first adult horror story, originally written in 1940, when he was a teenaged apprentice journalist.]

G51 THE GREAT VORE

a. _________, see Not After Nightfall: Stories of the Strange and the Terrible [B6].

[Note: This novella, another entitled 'The Restless Dead' (I8) and a third tale, all featuring the same character, were intended to comprise a volume for Arkham House entitled *Three of a Kind: Leaves from the Casebook of Professor Alden Kane.*]

b. _________, see Here Be Daemons: Tales of Horror and the Uneasy [B4].

G52 THE GREY HOUSE

a. _________, see Not After Nightfall: Stories of the Strange and the Terrible [B6].

b. _________, see From Evil's Pillow [B3].

c. _________, see Voices of Doom: Tales of Terror and the Uncanny [B8].

d. _________, in *The Mammoth Book of Zombies* edited by Stephen Jones, Robinson Publishing, London, England, 1993 (tp).

e. _________, in *The Mammoth Book of Zombies* edited by Stephen Jones, Carroll & Graf, New York, New York, 1993 (tp).

f. _________, in *The Giant Book of Zombies* edited by Stephen Jones, Magpie Books, London, England, 1995 (tp).

g. _________, in *The Giant Book of Zombies* edited by Stephen Jones, The Book Company, Sydney, Australia, 1995 (tp).

h. _________, in *The Giant Book of Zombies* edited by Stephen Jones, Parragon, London, England, 1995 (tp).

i. _________, in *The Mammoth Book of Haunted House Stories* edited by Peter Haining, Robinson, London, England, 2000 (tp).

j. _________, in *The Mammoth Book of Haunted House Stories* edited by Peter Haining, Carroll & Graf, New York, New York, 2000 (tp)*.

k. _________, in *The Mammoth Book of Haunted House Stories* edited by Peter Haining, Constable & Robinson Ltd, London, England, 2005 (tp).

[NOTE: This is a revised version of the 2000 edition.]

l. _________, in *The Mammoth Book of Haunted House Stories* edited by Peter Haining, Carroll & Graf, New York, New York, 2005 (tp)*.

[NOTE: This is a revised version of the 2000 edition.]

(*"The novella, 'The Grey House', stemmed from the atmosphere of a 15th-century house in France my mother-in-law purchased and restored. I first saw it as a ruin, but its terrace and outlook down to an equally ancient park, a ruined summerhouse and the river winding beyond, quickly gave me the physical aspects of the story though, as many readers know, I transformed the innocuous park into a graveyard from which the miasmas and manifestations were to come."* BC.)

G53 THE HOUSE BY THE TARN

a. __________, in *Dark Things* edited by August Derleth, Arkham House: Publishers, Sauk City, 1971 (hc).
(*"I discovered Arkham House during the first two or three years of the war, when rare copies found their way to British shores. Many years later, when I was establishing myself as a professional author of novels, non-fiction studies and macabre stories, on impulse I wrote to August Derleth. He replied in most cordial terms, asking to see my work. Some of my stories had already seen publication in Pan books and I was astonished to learn he had already heard of me. His encouragement, long and enjoyable letters and his patronage when he offered to publish my first American hardback under the imprint of his famous Arkham House, formed one of the warmest and most delightful strands of my writing career."* BC.)

b. __________, in *Spectre 4:* edited by Richard Davis, Abelard Schuman Limited, London, England, 1977 (hc).

c. __________, see Voices of Doom: Tales of Terror and the Uncanny [B8].

d. __________, in *The Mammoth Book of Ghost Stories* edited by Richard Dalby, Robinson Publishing, London, England, July 1990 (tp).

e. __________, in *The Mammoth Book of Ghost Stories* edited by Richard Dalby, Carroll & Graf, New York, New York, 1990 (tp)*.

f. __________, in *The Anthology of Ghost Stories* edited by Richard Dalby, Tiger Books, England, 1994 (hc)*.

g. __________, in *Phantastic Book of Ghost Stories* edited by Richard Dalby, Barnes & Noble Books, New York, New York, 1996 (hc)*.

h. __________, in *Phantastic Book of Ghost Stories* edited by Richard Dalby, MetroBooks, New York, New York, 2002 (hc)*.

i. __________, in *The World's Greatest Ghost Stories* edited by Richard Dalby, Magpie Books, London, England, 2004 (tp)*.

j. __________, in *The World's Greatest Ghost Stories* edited by Richard Dalby, Barnes & Noble Books, New York, New York, 2004 (tp)*.

k. __________, in *The World's Greatest Ghost Stories* edited by Richard Dalby, Magpie Books, London, England, 2005 (hc)*.
l. __________, in *The World's Greatest Ghost Stories* edited by Richard Dalby, Barnes & Noble Books, New York, New York, 2005 (hc)*.

G54 HUNTED BY WOLVES

a. __________, see Knife in the Back: Tales of Twilight and Torment [B5].

G55 ILL MET BY DAYLIGHT

a. __________, see Cold Hand on My Shoulder: Tales of Terror & Suspense [B2].
b. __________, in *The Mammoth Book of Best New Horror Volume Fourteen* edited by Stephen Jones, Robinson, London, England, 2003 (tp).
c. __________, in *The Mammoth Book of Best New Horror Volume Fourteen* edited by Stephen Jones, Carroll & Graf, New York, New York, 2003 (tp).

(*"I must admit that my work in certain genres has been greatly influenced by the wonderful tales of M.R. James. His own subtle and oblique way of constructing narrative has been my dictum: less is more. I also admire the work of the distinguished film director Lawrence Gordon Clark, who has translated some of James' finest stories to the television screen in a restrained and wonderfully evocative fashion. Clark has been trying to get my film treatment of James' 'Count Magnus' made for TV, for which I shall always be grateful.*

"I hope that those who come to this tale for the first time will find some measure of enjoyment and, dare one say it, a frisson *of the terror that takes hold of the protagonist. For, as Edward van Sloan so memorably observed in an epilogue to one of the classic Universal horror films, 'There are such things!'"* BC.)

G56 IN A DARKLING WOOD

a. __________, see Whispers in the Night: Stories of the Mysterious and Macabre [B9].

G57 THE JANISSARIES OF EMILION

a. __________, in *The Eighth Pan Book of Horror Stories* edited by Herbert Van Thal, Pan Books, London, England, 1967 (pb).
b. __________, see Not After Nightfall: Stories of the Strange and the Terrible [B6].

c. __________, in *Alfred Hitchcock Presents: A Month of Mystery* [edited by Robert Arthur], Random House, New York, New York, 1969 (hc).
d. __________, in *Alfred Hitchcock Presents A Month of Mystery* [edited by Robert Arthur], Max Reinhardt, London, England, 1969 (hc).
e. [as **JANITSHARERNE FRA EMILION**], in *Hitchcock Præsenterer Uhyggelige uger*, Steen Hasselbalchs Forlag, Vojens, Denmark, 1970 (pb).
f. [as **KILLERS FROM THE PAST**], in *Reveille*, London, England, Week ending November 21, 1970 (newspaper).
g. __________, in *Alfred Hitchcock Presents: Dates with Death: Tales from A Month of Mystery* edited by Alfred Hitchcock [and Robert Arthur], Dell Books, New York, New York, January 1972 (pb).
h. __________, in *Alfred Hitchcock Presents A Month of Mystery Book Two*, Pan Books, London, England, 1973 (pb).
i. __________, see When Footsteps Echo: Tales of Terror and the Unknown [B7].
j. __________, see And Afterward the Dark: Seven Tales [B1].

G58 **JUSTICE AT THE CROSSROADS**

a. __________, see Here Be Daemons: Tales of Horror and the Uneasy [B4].
[Note: This story was a revised version of the opening chapter of *The Vampire: In Legend, Fact and Art* (F21).]

G59 **THE KNOCKER AT THE PORTICO**

a. __________, in *Dark Things* edited by August Derleth, Arkham House: Publishers, Sauk City, 1971 (hc).

(*"I corresponded with August Derleth for some five years, regarded him as a friend and had great affection for him; his efforts helped me to get established as a macabre and fantasy writer in the United States. Kind, generous–often paying his authors in advance for material he knew he could not use until years later. He was about the last of the all-around literateurs, excelling in so many genres; but his cruel workload, which he carried on seven days a week, working all his life against sixty-day promissory notes, eventually undermined even his iron constitution."* BC.)

b. __________, in *The Year's Best Horror Stories No.2* edited by Richard Davis, Sphere Books, London, England, 1972 (pb).
[Note: "Foreword by Christopher Lee".]
c. __________, in *The Year's Best Horror Stories: Series II* edited by Richard Davis, DAW Books Inc., New York, New York, July 1974 (pb).
[Note: "Foreword by Christopher Lee".]

d. __________, see Here Be Daemons: Tales of Horror and the Uneasy [B4].

e. [as **DES COUPS À LA PORTE**], in *Trilogie ténébreuse* edited by Richard D. Nolane, Olivier Raynaud/Presses du Créépuscule, Aix-en-Provence, France, June 1981.

[Note: This book (*Dark Trilogy*), which was also supposed to include stories by Ramsey Campbell and Brian Lumley, was apparently never published.]

f. __________, in *12 Gothic Tales* edited by Richard Dalby, Oxford University Press, Oxford, England/New York, New York, 1998 (tp).

[Note: Titled *Twelve Gothic Tales* on the cover.]

g. [as **UKSETAGUNE KOPUTAJA**], in *Sünged Varjud 2: Saatana Teener* edited by Koostanud Raul Sulbi, Kirjastus Fantasia, Tartu, Estonia, 2004 (hc).

G60 **LINE ENGAGED**

a. __________, in *Chronicles of Crime: The Second Ellis Peters Memorial Anthology of Historical Crime* edited by Maxim Jakubowski, Headline Book Publishing, London, England, 1999 (tp).

(*"Henry Ford once famously declared, 'History is bunk!' He may have been right in some respects; many historians have twisted the facts to fit their own pet theories regarding events which occurred long ago, but there is no doubt that history comes vividly alive in the best fiction devoted to that genre.*

"From an early age I was gripped by Conan Doyle's historical novels, notably Sir Nigel *and* The White Company *and, of course, his Sherlock Holmes tales. Mary Shelley's* Frankenstein *and such Gothic novels as* The Monk *and* The Castle of Otranto *certainly influenced my own work, both in my Gothic novels,* The Black Death, Necropolis *and* The House of the Wolf, *and my Sherlockian pastiches, the Solar Pons series. These works were set in my favourite periods for fictional recreations of the past, the late nineteenth century and the 1920s. These periods embraced most of my fictional favourites in the criminal genre; Sheridan Le Fanu, H.G. Wells–whose horrific short stories are often overlooked today–Robert Louis Stevenson, Poe, M.R. James and E.F. Benson, particularly. The fog-bound streets of London once haunted by Jack the Ripper, bleak moorlands and old houses whose very stones breathe the stuff of history, enchant the reader safely ensconced in an armchair in the security of a modern well-lit room.*

"My present tale, though bringing history up to date in that it is set in the present, borrows all those historical aspects of the Gothic tale while remaining firmly within the parameters of the genteel criminal genre, the very ordinariness of which, in real life, adds to the horror. Or at least, one hopes so." BC.)

b. __________, in *Chronicles of Crime: The Second Ellis Peters Memorial Anthology of Historical Crime* edited by Maxim Jakubowski, Headline Book Publishing, London, England, 2000 (pb).

b. __________, see Cold Hand on My Shoulder: Tales of Terror & Suspense [B2].

G61 **THE MADONNA OF THE FOUR-ALE BAR**

a. __________, see Voices of Doom: Tales of Terror and the Uncanny [B8].

b. __________, in *John Creasey's Crime Collection 1990: An Anthology by Members of the Crime Writer's Association* edited by Herbert Harris, Victor Gollancz, London, England, 1990 (hc).

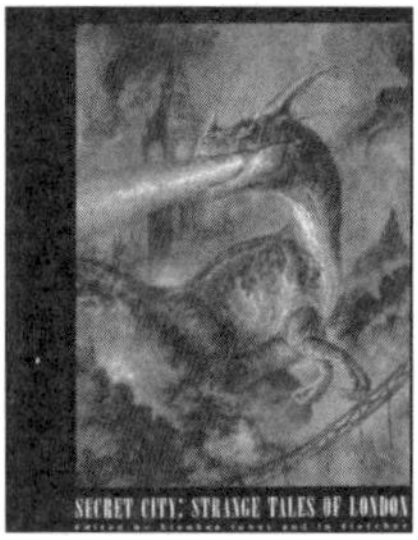

c. __________, in *Secret City: Strange Tales of London* edited by Stephen Jones and Jo Fletcher, The 1997 World Fantasy Convention/Titan Books, London, England, October 1997 (convention book).

d. __________, in *Secret City: Strange Tales of London* edited by Stephen Jones and Jo Fletcher, The 1997 World Fantasy Convention/Titan Books, London, England, October 1997 (hc). [Note: Published in a 300-copy signed edition.]

G62 **A MESSAGE FROM THE STARS**

a. __________, in *Space 2: A Collection of Science Fiction Stories* edited by Richard Davis, Abelard-Schuman , London, England, 1974 (hc). [Note: Published in this volume as 'Message from the Stars'.]

b. __________, see When Footsteps Echo: Tales of Terror and the Unknown [B7].

G63 **MRS. VAN DONK**

a. __________, see Here Be Daemons: Tales of Horror and the Uneasy [B4].

b. __________, in *The Urbanite* No.9: *Strange Places*, Fall 1997, edited by Mark McLaughlin, Urban Legend Press, Davenport, USA (magazine). [Note: "Featured Writer: Basil Copper"].

G64 **MURDER AT THE ZOO**

a. __________, see Solar Pons #11: The Uncollected Cases of Solar Pons [C11].

b. __________, see The Further Adventures of Solar Pons [C5].

G65 **THE OBELISK**

a. __________, see Whispers in the Night: Stories of the Mysterious and Macabre [B9].

G66 OLD MRS. CARTWRIGHT

a. __________, see Not After Nightfall: Stories of the Strange and the Terrible [B6].
b. __________, see Here Be Daemons: Tales of Horror and the Uneasy [B4].
c. __________, in *100 Fiendish Little Frightmares* edited by Stefan Dziemianowicz, Robert Weinberg and Martin H. Greenberg, Barnes & Noble Books, New York, New York, October 1997 (hc).

G67 ONE FOR THE POT

a. __________, see Whispers in the Night: Stories of the Mysterious and Macabre [B9].

G68 OUT OF THE FOG

a. __________, in *Argosy* Vol.XXXI, No.9, England, September 1970 (magazine).
b. __________, in *John Creasey's Mystery Bedside Book 1971: The Anthology of the Crime Writers Association* edited by Herbert Harris, Hodder and Stoughton/King Crime, 1970 (hc).
c. [as **I TAGEN**], in *Dansk Familieblad*, Number 33, August 13 1971, Copenhagen, Denmark (magazine).
d. __________, see When Footsteps Echo: Tales of Terror and the Unknown [B7].
e. __________, in *Fantasycon XX Programme Book* edited by Peter Coleborn, Fantasycon XX, Acocks Green, Birmingham, 1996 (convention book).
f. __________, in *The Unexplained: Stories of the Paranormal* edited by Ric Alexander [Peter Haining], Orion Books, London, England, 1998 (hc).
g. __________, in *The Unexplained: Stories of the Paranormal* edited by Ric Alexander [Peter Haining], Orion Books, London, England, 1998 (tp)*.

G69 OUT THERE

a. __________, see Whispers in the Night: Stories of the Mysterious and Macabre [B9].

G70 POETIC JUSTICE

a. __________, see Cold Hand on My Shoulder: Tales of Terror & Suspense [B2].

G71 PROFESSOR MAHMOUD

a. __________, see this volume.

G72 **QUEEN BEE**

a. __________, in *Don't Turn Out the Light* edited by Stephen Jones, PS Publishing, Harrogate, England, December 2004 (hc). There are two varients of this edition:

i. Description: 500-copy edition signed by the editor.

ii. Description: 200-copy slipcased deluxe edition signed by all the living contributors.

(*"The genesis of this story is based on a number of things, all connected with insects. I do a good deal of writing during my two long breaks I spend with my wife at her family home in France. I work in the mediaeval part of the house, where there are large skylight windows in the slanting roof. One afternoon, some years ago, I was slaving away at my typewriter when a great hornet flew in, making a droning noise like a World War II bomber.*

"Their sting can cause death. This one was several inches long and obviously angry. My wife and I then began an epic battle, with myself wielding a rolled newspaper as it dive-bombed me. I eventually knocked it to the floor and, by putting layers of newspaper over it, I actually had to stand on it to kill it. There was a terrific crack as the armour-like body collapsed.

"I hate killing any living creature, but it had to be done as hornets are deadly, as I have already mentioned.

"Some years ago something even worse happened at our house in Kent. There were three wasp's nests in the loft and one night the creatures actually gnawed through the wall of our bedroom and a huge swarm descended on us at 4:00 in the morning. It was like something out of a horror film as we fled, slamming the door behind us. When the nests had been destroyed, we had to have metal plates fitted behind the wall, as the wasps seem to like our house, returning again and again.

"Some of these things were at the back of my mind when I was working at our house in France in May 2004 and a bee flew in at the open window. They are generally benevolent creatures, but its persistent buzzing got on my nerves. It eventually flew off, but it had set me thinking. I took the story I was working on out of my typewriter and sat down at my writing table with a fresh sheet in the machine.

"I wrote 'Queen Bee' in an hour, and I hope readers will get a taste of the horror I felt during my battle with the hornet. Just lately a bee has been coming through the French windows in our dining room. Nature imitating art, or vice versa?" BC.)

G73 **RAP-SHEET**

a. __________, see this volume.

[Note: Opening fragment of a proposed new Mike Faraday novel (#53).]

G74 **READER, I BURIED HIM!**

a. __________, in *The Vampire Omnibus* edited by Peter Haining, Orion Books, London, England, 1995 (hc).

b. __________, in *The Vampire Omnibus* edited by Peter Haining, Orion Books, London, England, 1995 (tp)*.

c. __________, in *The Vampire Omnibus* edited by Peter Haining, Orion Books, London, England, 1996 (pb).

d. __________, see Whispers in the Night: Stories of the Mysterious and Macabre [B9].

e. __________, in *The Vampire Omnibus* edited by Peter Haining, Kadokawa bunko, Japan, March 1999*.

f. [as **... UND DANN BEGRUB ICH IHN**], in *H.R. Giger's Vampirric* edited by H.R. Giger, Festa-Verlag, Leipzig, Germany, May 2003 (tp)*.

(*"In the autumn of 1993, while I was in France, there was an extraordinary item on French TV. Dozens of sheep had been found dead by shepherds and farmers in a wild and remote mountainous area of the Var. The sheep had been attacked during the night as they grazed by something which crept beneath them and with sucker-like mouth had drained all their blood.*

"In front of the TV cameras, a farmer held up one sheep which had survived the attack and there was a large hole in its stomach about the size of a 50 pence coin. The police were baffled–as they say–and medical experts and scientists, after careful examination of the area and the dead animals, were equally mystified. So far as they knew, there was no such creature known to science as the one which had killed the sheep. I must confess my hair stood up on end!" BC.)

G75 **THE RECOMPENSING OF ALBANO PIZAR**

a. __________, see From Evil's Pillow [B3]

b. __________, in *The Year's Best Horror Stories No.3* edited by Richard Davis, Sphere Books, London, England, 1973 (pb).

c. [as **WER DIE WAHL HAT, HAT DIE QUAL**], in *Der Totenvogel: Sechs Gruselstories* edited by Richard Davis, Erich Pabel Verlag KG/Vampir-Taschenbuch, Rastatt, West Germany, July 1975 (pb).

d. __________, see When Footsteps Echo: Tales of Terror and the Unknown [B7].

G76 **REFLECTIONS**

a. __________, see Knife in the Back: Tales of Twilight and Torment [B5].

G77 **RIDING THE CHARIOT**

a. __________, see Whispers in the Night: Stories of the Mysterious and Macabre [B9].

G78 **THE SEALED SPIRE MYSTERY**

a. __________, see Solar Pons #8: The Dossier of Solar Pons [C2a].
b. __________, see The Dossier of Solar Pons [C2b].

G79 THE SECOND PASSENGER

a. __________, in *The Dark Brotherhood Journal 3*, Vol.2 No.1, July 1973 (magazine).
b. __________, see Here Be Daemons: Tales of Horror and the Uneasy [B4].
c. __________, in *The Mammoth Book of Ghost Stories 2* edited by Richard Dalby, Robinson Publishing, London, England, 1991 (tp).
d. __________, in *The Mammoth Book of Ghost Stories 2* edited by Richard Dalby, Carroll & Graf, New York, New York, 1991 (tp)*.
e. __________, in *The Giant Book of Ghost Stories* edited by Richard Dalby, Magpie Books Ltd., London, England, 1993 (tp)*.

G80 **SHAFT NUMBER 247**

a. __________, in *New Tales of the Cthulhu Mythos* edited by Ramsey Campbell, Arkham House, Sauk City, Wisconsin, 1980 (hc).
b. __________, in *New Tales of the Cthulhu Mythos Vol.1* edited by Ramsey Campbell, Kokusho-Kankohkai, Japan, September 1983*.
c. __________, in *New Tales of the Cthulhu Mythos* edited by Ramsey Campbell, Grafton Books, London, England, 1988 (pb).
d. [as **LE PUITS NUMÉRO 247**], in *Le Livre noir: nouvelles légendes du mythe de Cthulhu* edited by Ramsey Campbell, Presses Pocket, Paris, France, April 1991 (pb)*.
e. __________, in *Cthulhu 2000: A Lovecraftian Anthology* edited by Jim Turner, Arkham House Publishers, Sauk City, Wisconcin, 1995 (hc).
f. __________, in *Cthulhu 2000: A Lovecraftian Anthology* edited by Jim Turner, Dey Rey, New York, June 1999 (pb)*.
g. [as **IL POZO NUMERO 247**], in *Cthulhu 2000* edited by Jim Turner, Sperling &

Kupfer Editori, Milan, Italy, February 1999 (hc).

h. [as **EL POZO NÚMERO 247**], in *Cthulhu 2000* edited by Jim Turner, La Factoriía de Ideas/Solaris Fantasía, Madrid, Spain, 2000 (tp)*.

i. [as **SACHTA CÍSLO 247**], in *Cthulhu 2000* edited by Jim Turner, Polaris, Czech Republic, 2001 (tp)*.

j. [as **EL POZO NÚMERO 247**], in *Cthulhu 2000* edited by Jim Turner, La Factoría de Ideas/Solaris Fantasía, Madrid, Spain, 2003 (tp)*.

k. [as **LA SECCIÓN 247**], in *Nuevos cuentos des los mitos de Cthulhu* edited by Ramsey Campbell, Valdemar Gótica, Spain, 2003 (tp)*.

l. [as **SCHACHT NUMMER 247**], *Spur de Schatten: Neue Geschichten aus dem Cthulhu-Mythos* edited by Jim Turner, Bastei Lübbe, Germany, January 2004 (tp)*.

(*"I rate Lovecraft very highly indeed and at his very best the next in succession to the great 19th century master. Poe is, of course, now recognised as a classic author both in prose and poetry and, though in a somewhat lower niche, I am certain that Lovecraft will in time–though it may take another thirty years–take his true place as a classic writer in the field he made uniquely his own."* BC.)

G81 **THE SILVER SALAMANDER**

a. __________, see Knife in the Back: Tales of Twilight and Torment [B5].

G82 **THE SPIDER**

a. __________, in *The Fifth Pan Book of Horror Stories* edited by Herbert Van Thal, Pan Books, London, England, 1964 (pb).

b. __________, see Not After Nightfall: Stories of the Strange and the Terrible [B6].

c. __________, see And Afterward the Dark: Seven Tales [B1].

d. __________, in *100 Fiendish Little Frightmares* edited by Stefan Dziemianowicz, Robert Weinberg and Martin H. Greenberg, Barnes & Noble Books, 1997 (hc).

[Note: Editor Helen Hoke wanted to reprint this story in her 1978 YA anthology, *Terrors, Torments and Traumas*, but for some reason the contracts were never signed and returned.]

(*"Macabre story ideas seem to come from the subconscious and then rapidly crystallise from some external event. 'The Spider', for instance, arose from our discovery of an enormous brown spider on a bedroom wall when I was on holiday in France many years ago."* BC.)

G83 STORM OVER STROMJOLLY

a. __________, see Knife in the Back: Tales of Twilight and Torment [B5].

G84 THE STRANGER

a. __________, see Voices of Doom: Tales of Terror and the Uncanny [B8].

G85 THE SUMMERHOUSE

a. __________, see Whispers in the Night: Stories of the Mysterious and Macabre [B9].

G86 THE TREASURE OF OUR LADY

a. __________, see Here Be Daemons: Tales of Horror and the Uneasy [B4].

G87 THE TRODES

a. __________, in *Armada Sci-Fi 1* edited by Richard Davis, Armada/William Collins, London, England, 1975 (pb).
b. __________, see Here Be Daemons: Tales of Horror and the Uneasy [B4].

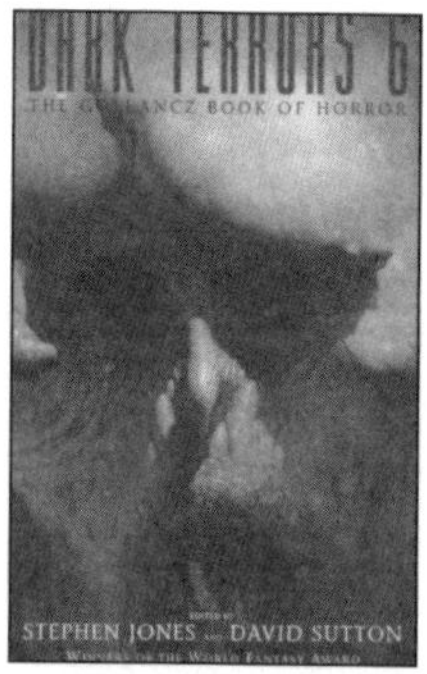

G88 THERE LIES THE DANGER . . .

a. __________, in *Dark Terrors 6: The Gollancz Book of Horror* edited by Stephen Jones and David Sutton, Victor Gollancz/ Orion Books, London, England, 2002 (hc).
b. __________, in *Dark Terrors 6: The Gollancz Book of Horror* edited by Stephen Jones and David Sutton, Victor Gollancz/Orion Books, London, England, 2002 (tp).
c. __________, in *Dark Terrors 6: The Gollancz Book of Horror* edited by Stephen Jones and David Sutton, Victor Gollancz/Orion Books, London, England, 2004 (tp).
d. __________, see Knife in the Back: Tales of Twilight and Torment [B5].

G89 **A VERY PLEASANT FELLOW**

a. __________, see From Evil's Pillow [B3].

b. __________, see Voices of Doom: Tales of Terror and the Uncanny [B8].

G90 **VOICES IN THE WATER**

a. __________, in *Weird Shadows Over Innsmouth* edited by Stephen Jones, Fedogan & Bremer, Minneapolis, Minnesota, 2005 (hc).

There are varients of this edition:

i. Description: limited edition.

ii. Description: 100-copy slipcased, numbered edition signed by all the living contributors.

G91 **THE WAY THE WORLD DIED**

a. __________, see Here Be Daemons: Tales of Horror and the Uneasy [B4].

G92 **WHEN GREEK MEETS GREEK**

a. __________, in *The Mammoth Book of Dracula: Vampire Tales for the New Millennium* edited by Stephen Jones, Robinson, London, England, 1997 (tp).

b. __________, in *The Mammoth Book of Dracula: Vampire Tales for the New Millennium* edited by Stephen Jones, Carroll & Graf, New York, New York, 1997 (tp).

c. [as **QUANDO UN GRECO INCONTRA UN ALTRO GRECO**], in *Il Grande Libro di Dracula* edited by Stephen Jones, Newton & Compton Editori, Rome, Italy, 2000 (tp).

d. __________, see Cold Hand on My Shoulder: Tales of Terror & Suspense [B2].

e. [as **QUANDO UN GRECO INCONTRA UN ALTRO GRECO**], in *Il Grande Libro di Dracula* edited by Stephen Jones, Newton & Compton, Rome Editori, Italy, 2004 (tp).

G93 **THE WHITE TRAIN**

a. __________, see Knife in the Back: Tales of Twilight and Torment [B5].

G94 WISH YOU WERE HERE

a. __________, in *Horror for Christmas* edited by Richard Dalby, Michael O'Mara Books Limited, London, England, October 1992 (hc).

b. __________, in *Horror for Christmas* edited by Richard Dalby, Headline Book Publishing, London, England, October 1993 (tp)*.

c. __________, in *Mistletoe & Mayhem: Horrific Tales for the Holidays* edited by Richard Dalby, Book Sales Incorporated, New York, New York, 1993 (hc)*.

d. __________, in *Mistletoe & Mayhem: Horrific Tales for the Holidays* edited by Richard Dalby, Book Sales Incorporated, New York, New York, 1993 (pb)*.

e. __________, in *Mistletoe & Mayhem: Horrific Tales for the Holidays* edited by Richard Dalby, Castle Books, Seacaucus, New Jersey, 1993 (hc)*.

f. __________, in *Horror for Christmas* edited by Richard Dalby, Brockhampton Press/Caxton Publishing Group, London, England, December 1995 (hc)*.

g. __________, see Whispers in the Night: Stories of the Mysterious and Macabre [B9].

PETER HAINING: A TRIBUTE

I was dearly saddened to hear of the death of Peter Haining, the author and anthologist, at the age of sixty-seven, following a heart attack. Peter was a great friend and I first visited him at the offices of New English Library at Barnard's Inn, off Holborn in London.

The reason was that he published my first collection of macabre tales, *Not After Nightfall*, in paperback in 1967, for which he wrote an appreciative note. I shall always be grateful for the fine kick-start he gave me, and I am sure that other authors would concur.

As so often occurs, we lost contact and he then shifted his literary endeavours into various volumes on military history and aircraft. Much sympathy will be extended to his wife Philippa and family. We will not see his like again.

—Basil Copper

November, 2007

THE BLACK COFFIN

I

BRUCE MEREDITH had a bizarre recurring dream for three nights running, which was very unusual for him. A leading copywriter for a leading London advertising agency, he had a very smooth passage through life, where success had cone to him without any of the strenuous efforts that most people had to make in to survive in the face of the daily struggle. Perhaps things had been too easy, he thought; there had been many women in his life and there again his successes had been achieved without too much effort.

He was an extremely handsome man with great charm which many women found irresistible. And he had never ever dreamed before, except when he was a child. The thing had begun as a minor irritation which he had brushed aside as a momentary aberration on waking, but the second or third time the same images had assumed a darker, not to say sinister, aspect. It had begun in the most innocuous way. He was travelling somewhere–the destination was never clear and the details of the vehicle in which lie was being transported were equally vague.

At first the journey was pleasant, with sunlight streaming in at the compartment windows–if indeed that was the method of his conveyance–and he had glimpses of a delightful sylvan countryside which looked like something out of an old painting by Watteau.

by Basil Copper

To his disappointment, after awakening from the first dream his destination had not been revealed. This was bizarre in itself, as he could not have expected anything of that nature. But on this first occasion, as he bathed and dressed for another busy day at the office, the dream faded from his mind.

Despite his outward composure, Meredith was becoming desperate. The sylvan dreams be had experienced had turned darker and more terrifying as the days advanced. They featured a black coffin. Why this should be so he had no idea. It was probably his morbid imaginings, due to the difficult position he was in, entirely due to his profligate habits. He owed a good deal of money, mostly due to gambling debts, and was almost a month behind with his rent and the landlord was becoming more and more insistent. No wonder his sleeping hours were so sombre, he thought. Desperate situations called for desperate measures, and he was becoming ever more stressed as the days passed. The nights were becoming even worse, and he knew he had to make a huge effort to extricate himself from the morass into which he had descended.

The dream of the black coffin always returned after a broken sleep in the early hours, and he knew he had to get away from this small town with its parochial habits and start a new life elsewhere. The pale dawns when all colour was bleached from the sky were the worst. When he left his shabby room things were a little better. In the intervals from his menial job as a clerk in a local insurance office, he spent much time in the reading room of the local library, his twisted thoughts assessing useless solutions to all his problems.

Also, it was the one place where he was unlikely to be disturbed by creditors. After leaving the library, he walked warily and swiftly to a small cafe in a side street and bought a sandwich and a cup of coffee from his meagre savings. It was a half-day at the office, and he spent the afternoon walking, his mind uselessly assessing any possible solutions to his situation without coming to any successful conclusions. When he returned to his lodgings in the early dusk, his landlord told him that a man in dark clothes, looking like some sort of official, had been asking after him.

He owed so much money that it was a wonder that none of these gentlemen had appeared before, he thought.

He went up to his room in an abject condition and sat slumped in a chair. He eventually sought his bed and again dreamed of the black coffin floating in space.

He awoke at 6:00 a.m., a pale ray of sunshine caressing his tormented face. There was no other way out than the path he had suddenly resolved to follow. He quickly dressed and sat at the rickety table in the corner and wrote a detailed note with a stub of pencil.

Then he went to his suitcase and took out a leather satchel which contained a number of bottles and powders. He went to the washbasin, filled a tooth glass with water and stirred in a strong dose of the powder which he knew would do the trick. A few moments later he returned the material to the suitcase and then lay down on the bed to await his destiny.

II

Meredith's death caused a minor sensation in the small town where such things were a rarity. There was a hasty inquest and a formal verdict of suicide while of unsound mind, which greatly disappointed all his debtors.

But the circumstances that later followed were bizarre in the extreme.

There had bee a fire at the undertaker's following the inquiry, and the coffin containing Meredith's body had been charred and black. It was more like an incineration, people in the town thought. Meredith's visions in his dreams had been fulfilled. The coffin had been gliding down the Styx in Meredith's last dream, but the reality was very different. These and even stranger happenings were the source of much speculation in the local press. They would be the main topic of conversation among locals for many years to come.

"A very clever chap."

Two men were sitting in a local restaurant, thoughtfully stirring their coffee and discussing Meredith a few days later.

"He should have been in jail," Smithers said.

Thompson chuckled. "Hardly likely."

"There was nothing much in the papers," Smithers continued.

His companion shrugged. "Not surprising, under the circumstances."

"What do you mean?"

"It was extremely weird. After he was pronounced dead his body was conveyed to the mortuary while a pauper's funeral was prepared."

As Thompson momentarily broke off his narrative, Smithers took another sip of his coffee, giving his companion a puzzled look. "Nothing unusual in that, surely?" he said in answer to Thompson's last remark.

Thompson leaned across the table. "I had this from a friend in the C.I.D. It won't be in the papers. It's beyond belief."

"In what way?"

"Just this. The body was left in the coffin with the lid left unscrewed. Just in case any relative turned up with a wish to view the body. But in the night there was a fire in the mortuary chapel."

"Really?"

Thompson gave his friend a brief smile. "Remember, the chapel is a bleak place constructed entirely of stone and the building was securely locked. There was no combustible material there and, in any case, no means of starting a fire."

Thompson again lowered his voice to a whisper. "The coffin was a charred mass, absolutely black, as though a blow torch had been applied to it."

"So he was incinerated?"

Thompson shook his head. "There was nothing in the coffin. Just a handful of dust."

"You mean the body was completely consumed?"

Thompson had a remarkable expression on his face which his friend was unable to read. "Of course not. There was nothing there."

"Amazing."

"Not half as amazing as the situation facing the police and the civil authorities. So they decided to hush the whole thing up."

"I see."

"I'm glad you do. But there is a postscript to all this. He paused. "The local stationmaster, who drinks at the same bar that I frequent, told me that a man of Meredith's build but with a large black moustache, boarded an early morning train for London on the morning of the fire."

"I don't understand."

Thompson curbed his impatience. He produced a photocopy of an old poster from his wallet.

Smithers read:

THE GREAT VANCE

His death defying feats are beyond belief. His famous exploits include being buried in a coffin for several days in a cataleptic trance before being released unharmed.

Smithers was bewildered. "I don't see . . ."

Thompson began to laugh. "Perhaps Meredith had also studied escapology?"

Smithers was still nonplussed. "But how did he start the fire and then escape a locked room?"

Thompson laughed again. "Ah, there you have me . . . A trade secret, perhaps?"

H. MEDIA ADAPTATIONS

Includes an alphabetical listing of adaptations in other media based on work by Basil Copper

(*"The cinema of the golden age of the German silents and, of course, the Universal classics of the 1930s have been another strong influence, and I have built up a huge collection of films, over 1,000 titles in toto, over many decades."* BC.)

H1 CAMERA OBSCURA

a. _________, *Rod Serling's Night Gallery*, Universal/NBC-TV, USA, December 8, 1971. Colour.

Director: John Badham. Producer: Jack Laird. Teleplay by Rod Serling. With: Rene Auberjonois (William Sharsted), Ross Martin (Mr. Gingold), Milton Parsons (Old Lamplighter), Philip Kenneally (Driver), Brendan Dillon (Drucker), Arthur Malet (Abel Joyce), John Barclay (Sharsted, Sr.).

[Note: "We branch out a bit and get into the area of photography. Now, this painting here had best be viewed in a darkroom, because it conjures up the ghostly, the ghastly, and the ghoulish. It tells the story about a very remarkable device. Our painting is called 'Camera Obscura'." First UK screening: January 7, 1974. Another Basil Copper story, 'Doctor Porthos', was bought by Universal Studios for *Rod Serling's Night Gallery*, but it was never filmed.]

H2 THE CANDLE IN THE SKULL

a. _________, on *Short Story*, BBC Radio 4, England, October 31, 1997.

Producer: Louise Grealish. Read by Gerard McDermott.

[Note: "Martin's attempt to dispose of his wife does not go according to plan—but his daughter gets a sinister treat for Halloween." Fifteen-minute reading of Basil Copper's story.]

H3 **COUNT MAGNUS**

a. __________, see this volume.

[Note: Unproduced shooting script for television, based on the 1904 story by M.R. James. Written June 3-28, 1974, the final six pages were revised on October 7, 1974.]

H4 **THE FALL OF THE HOUSE OF USHER**

a. __________, Trinity Films, England, 1952/2000. Tinted. Silent.

Additional Photography/Editing/Screenplay/Producer/Director: Basil Copper. From the story by Edgar Allan Poe. With: John Lurcook (The Visitor), Gordon Anckorn (Roderick Usher), Tom Emery (The Doctor), Joan-Marie (The Lady Madeline), Jack Shean (the Manservant), Jean Wyse (The Housekeeper), and with the special contribution of Kevin Connealy (Dr. Sarnov).

[Note: Shot in 1952 on 9.5mm black and white film by a group of Kent enthusiasts under the direction of Basil Copper, who also produced and wrote the script, this amateur production of Poe's famous story was an extremely ambitious project, which attracted the attention of national magazines. Filmed at Lime Tree Studios, Sevenoaks, and on atmospheric locations provided by poet John Pudney and Lord Sackville of Knole. Production came to a halt when the group lost the use of the studio and the footage languished in cans for forty-six years before Basil Copper resurrected it and approximately 42 minutes was assembled on video with titles by Gerald McKee. Without the equipment to edit the remaining 30 minutes of footage, the film remains tantalisingly unfinished.]

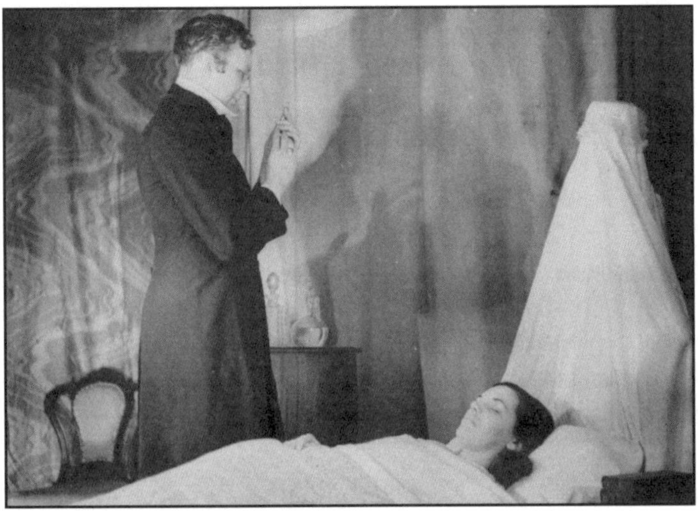

A scene from The Fall of the House of Usher

Left: The amateur film director in his cutting room

Right and below: Behind-the-scenes, The Fall of the House of Usher

H5 **INVITATION TO THE VAULTS**

R4 Revenge is the theme of tonight's **Fear on Four (11.00pm)**, adapted from Basil Copper's *Invitation to the Vaults*. The story reaches its gruesome climax in the dank vaults beneath the Palazzo Tortini where 'the tide floods in twice a day, and ahead of it the rats'.

Radio Times 85

a. __________, on *Fear on Four* (aka *The Man in Black*), BBC Radio 4, England, February 2, 1991.
Director: Martin Jenkins. Dramatised by Wally K. Daly. With: Ian Lindsay (Dr. Mazanares), Sara Badel (Madame Freitas), Roger Hammond (Albano Pizar), Emma Gregory (Secretary). Introduced by Edward de Souza (The Man in Black).
[Note: "The fourth in a series of eight nerve-tinglers . . . A crooked literary agent is subjected to a particularly harrowing revenge . . ." Adapted from Basil Copper's story 'The Recompensing of Albano Pizar', this was postponed from January 17, 1991 and became the last in the series to be broadcast.]
b. [as **ØYE FOR ØYE**], on *Radioteatret*, NRK Norwegian Broadcasting Corportion, Norway, November 22, 1992.
Intruksjon: Knut Thomassen. Teknisk regi: Bodil Moldsked. Dramatisert av Wally K. Daly, ööversatt av Helge Hagerup. With: Eva Bergh (Madame Freitas), Karl Bomann-Larsen (Albano Pizar).

H6 **PRESSURE POINT**

a. __________, Ulverscroft Large Print Books Ltd./Soundings Ltd., Leicester, England, May 1991.
Read by John Chancer.
[Note: Audio recording on four cassettes.]

Fedogan & Bremer publisher Philip J. Rahman and Basil Copper, London, England, circa late 1990s

COUNT MAGNUS

Adapted for television by BASIL COPPER

From the story by M.R. JAMES

CHARACTERS

MR. WRAXALL, an antiquarian.
COUNT DE LA GARDIE
NIELSEN, landlord of the inn.
HIS WIFE
PASTOR MOELLER
MAN IN THE CLOAK (Count Magnus)
HOODED FIGURE
MAN IN CLOAK (Pastor on boat)
HOODED FIGURE (Nun on boat)
COUNTRYMAN IN TRAIN
PASTOR (in flashback)
NIELSEN'S GRANDFATHER (in flashback)
HANS THORBJORN
ANDERS BJORNSEN
GIRL (in flashback)
THREE VILLAGERS (in flashback)
LANDLORD OF THE THREE CROWNS, SKARA
WOMAN WITH KEY AT THREE CROWNS
HOUSEKEEPER AT SKARA RECTORY
VILLAGERS, MEN DRINKING AT INN,
SERVANTS etc.

LOCATIONS

Exterior

Lake with steamer.
On board steamer.
Raback Manor House and grounds.
Countryside nearby.
Raback Church and Mausoleum.
Woods of Count Magnus' estate.
Raback station and platform.
Skara station and platform.
The Inn at Raback.
Countryside.
Lonely crossroads.
Streets of Skara town.
Three Crowns Inn, Skara.
The Parsonage, Skara.
Rooftops surrounding Three Crowns.
Lonely road.

Interior

Study, Raback Manor House.
Library, Raback Manor House.
Raback Church.
Mausoleum.
Inn at Raback: entrance hall; parlour; Wraxall's bedroom; Rooms differently dressed for flashbacks. Balcony at inn.
Three Crowns Inn, Skara: entrance hall; staircase; corridor; Wraxall's room.
Skara Church.
Railway Carriage

EXT. LAKE WITH STEAMER.

Fade in.

L.S. STEAMER approaching land.

EXT. COUNTRYSIDE.

M.S. A MIDDLE-AGED MAN in twenties clothes, carrying a BAG walks up to the doors of an INN.

C.S. WRAXALL looks up at a CHURCH SPIRE and pencils a note.

M.S. He walks along a path to a SIGNPOST in Swedish.

C.U. Dissolve to English translation: RABACK MANOR HOUSE.

L.S. A long, low white MANOR HOUSE. Dissolve to:

INT. STUDY, RABACK MANOR HOUSE.

An elegant study with white walls in the ancient manor house; thousands of BOOKS; a GLOBE of the world; a BIG FIREPLACE; OIL PAINTINGS; MR. WRAXALL is sitting in front of a DESK. Behind the desk is COUNT DE LA GARDIE, a thin, distinguished elderly man with white hair.

NARRATOR (V.O.)

In the spring of 1921 a Mr. Wraxall was travelling to Sweden to write a guide book. A scholar of Oxford, he was both intelligent and cultivated. A pleasant companion this fellow passengers and to the people encountered in his travels, he had only one fault perhaps. That was over inquisitiveness. Not a serious fault for a writer of guide-books. But to one who meddles in dark and curious places, it is a trait which might lead to serious and unpleasant consequences. Certainly it was a minor shortcoming for which Mr. Wraxall paid dearly enough in the end.

DE LA GARDIE

Certainly, my dear Mr. Wraxall, I would be only too happy to lend my assistance to your research. But Raback is a small place. I fear you will not find much to interest you here.

WRAXALL

On the contrary, Count, it promises to be fascinating for my purpose.

Reverse angle, favouring Wraxall. He is studying the room intently as the conversation continues.

DE LA GARDIE

I would be happy to accommodate you at the manor house.

WRAXALL

(in some embarrassment)

It is extremely kind of you, but I have already made arrangements to stay at the inn. But I would be grateful for the use of your library. I should make daily visits if that would not be inconvenient.

DE LA GARDIE

I have no wish to upset your itinerary. And the contents of my library is at your disposal. I am disappointed that you are not to partake of our hospitality.

WRAXALL

You are too kind. I should not presume to impose.

DE LA GARDIE

I would have been glad of your presence.

WRAXALL

I don't imagine you would have found a dull fellow like me much company.

Change angle. Wraxall with the Count seated at his desk in the background.

DE LE GARDIE

On the contrary, Mr. Wraxall. I take a keen interest in antiquities. Our church here, for instance... I have written a paper on its fabric and ornamentation.

WRAXALL

(brightening)

Really! I should like to see it.

DE LA GARDIE

(laughing)

You see, Mr. Wraxall, it would not have been so dull here as you feared! And your work is not unknown, even in Sweden. I have your book on Brittany.

C.S. Wraxall is obviously pleased. He smiles.

WRAXALL

You are too kind. But I must take my leave; I have imposed too long. And my supper waits at the inn. With your permission I will call early tomorrow afternoon. Perhaps we could visit the church together?

DE LA GARDIE

That would be most agreeable. I will try to find my paper by then.

De La Gardie stands up. Wraxall shakes hands with him. They go out the REAR DOOR as the camera tracks back to embrace the entire room.

Fade.

EXT. COUNTRYSIDE NEARBY.

Wraxall walks through the WALLED GARDEN of Raback Manor House. It is nearly dark and the wind is blowing the TREES mournfully.

Wraxall climbs a STEEP KNOLL, limned against the SUNSET. He gains the plateau on which the CHURCH stands.

The camera tracks with him among the PINE TREES.

He passes the curious church, with its WHITE WALLS, NARROW WINDOWS, and STEEP SPIRE. At one end is the MAUSOLEUM built by the original owners of the Manor House, and which has a DOMED ROOF with a PUMPKIN-SHAPED SPIRE on top. The dome is BLACK in contrast with the white walls.

The camera glides past the church.

C.U. Wraxall's face.

L.S. Tracking shot. A TALL FIGURE appears to be opening the DOOR of the mausoleum before the angle of the building blocks it from sight.

C.S. Wraxall's slightly puzzled face.

L.S. He walks up to the door of the inn.

Fade.

INT. INN AT RABACK—PARLOUR.

The inn parlour. A large, cheerful room with a BRIGHT FIRE in a BIG HEARTH, OAK PANELLING and lots of HEAVY, DARK FURNITURE. The walls white above the panelling, which is about eight feet high. A FEW PEOPLE are sitting drinking in the background. Wraxall sits at a table in the foreground eating his supper. A GIRL waits on him. The LANDLORD, a broad, burly man of middle-age called NIELSEN smokes a PIPE near the fire and close to Wraxall.

NIELSEN

The De La Gardies are much respected round about. Fine folk, with an ancient history in these parts. If facts are what you're after, you'll have enough for two books.

Wraxall smiles. He lifts his FLAGON OF WINE with satisfaction and after drinking puts it back on the table. He carries on the conversation between bites of FOOD.

WRAXALL

Perhaps you could help me. I need some specialised knowledge that only someone like you could give.

NIELSEN

Happy to oblige, sir, if I know just what you're after.

WRAXALL

Little things, that only local people would know. Curious stories, perhaps, or small quirks of regional history known only to people who live hereabouts.

Closer shot of Nielsen by fire. He puffs out smoke and looks shrewdly at Wraxall.

NIELSEN

You had something in mind?

C.S. Wraxall.

WRAXALL

It was just the atmosphere of Raback. It intrigues me. The church for instance. That mausoleum at the side . . .

He turns as they are interrupted by a CLATTER. The landlord's WIFE has dropped a TRAY in the background. He gets up to admonish her and to help clear up.

The landlord hurries back to Wraxall's side.

NIELSEN

You'll excuse me, sir. I'm wanted in the other bar.

WRAXALL

Perhaps we can take this up tomorrow?

Nielsen's attitude has subtly changed.

NIELSEN

We'll see, sir.

He hurries off. The girl comes out and starts MOPPING the floor.

C.S. Wraxall looks puzzled. He is about to say something to the serving girl but then changes his mind. He turns back to his food.

Fade.

EXT. RABACK CHURCH AND MAUSOLEUM.

Fade in.

L.S. The knoll and the church. Wraxall and De La Gardie are two minute SILHOUETTES on the skyline. They are strolling toward the church.

Closer shot.

WRAXALL

I must say, Count, your researches have been both thorough and scholarly. I am most impressed.

(he waves a THICK BUNDLE OF MANUSCRIPT)

These will be invaluable.

Favouring the Count.

DE LA GARDIE

I am glad to have been of help. You may keep the material as I have other copies.

Tracking shot as they walk toward the church. Favouring Count.

DE LA GARDIE (CONT'D)

You will be staying long in the neighbourhood?

WRAXALL

About four days I should think. There are many historic buildings hereabout. And I like to write up my notes and then re-check the site. But here we are.

The couple walk up to the BIG DOOR of the CHURCH PORCH. Wraxall stands back to allow the Count to enter ahead of him.

INT. RABACK CHURCH.

Church interior, camera shooting toward the door as the couple come in, their feet echoing over the FLAGS. Dark PEWS and GALLERIES; PULPIT covered with PAINTED CHERUBS and SAINTS. They start looking around.

ECHOES under the ceiling as they continue the tour. Wraxall carefully examines an ELABORATELY PAINTED ORGAN.

WRAXALL

You mention this specifically, I believe.

(he consults the Count's NOTES while he stands by and watches. He turns back to De La Gardie)

Remarkably thorough.

The Count smiles with satisfaction. Wraxall runs his fingers with almost sensual pleasure over the CARVINGS.

The couple walk up STAIRS into a GALLERY, the camera at the head of the stairs. Wraxall looks about him and consults the notes from time to time. He produces a PAD AND PENCIL and jots down his own impressions.

Wraxall looks up at the remarkable PAINTED CEILING.

The camera pans slowly over the details of THE LAST JUDGEMENT, lingering on BURNING SOULS, SMILING DEMONS, CITIES and SHIPS in FLAMES.

The camera pans slowly down to Wraxall and Count as they continue to look up.

C.S. Wraxall, who moistens his lips and glances at his companion.

WRAXALL

(enthusiastically)

This really is worth the visit. It gives a wonderful impression of that simplistic view of the world. It was all so black and white, then.

(he consults the notes)

Seventeenth-century, you say?

De La Gardie nods and takes his gaze off the ceiling.

DE LA GARDIE

An unknown artist. Though my family believes he belonged to a distant branch of my family.

Wraxall nods.

WRAXALL

Your family has been at the Manor House for 200 years, I understand. You inherited from an even older family.

Two-shot. A faint shadow passes over the Count's face.

DE LA GARDIE

That is correct. We inherited by marriage.

He breaks off and walks a little way down the gallery, leaving Wraxall standing.

DE LA GARDIE (CONT'D)

Have you seen these carvings?

Wraxall follows him.

GROUND FLOOR of the church. The couple walk back, stopping in leisurely fashion to examine VARIOUS FEATURES. They halt in an angle in which there is a CLEAR WINDOW. Through it can be seen part of the mausoleum building outside. Wraxall looks puzzled The Count follows his glance.

WRAXALL

Can we obtain access to this other building? It look equally interesting.

(he glances at the notes again)

You say little about it here.

De La Gardie bites his lip and moves awkwardly away.

DE LA GARDIE

The mausoleum? I believe you will find a reference there. It belonged to the original family who built the Manor House. There is nothing of any great note.

Wraxall looks at him sharply, adjusts his SPECTACLES.

WRAXALL

All the same, Count, I should very much like to see inside.

C.S. De La Gardie. A faint expression of annoyance crosses his face.

DE LA GARDIE

Then you will have to apply to the rector, for he alone has the keys.

He moves off, seemingly having lost interest. Wraxall looks toward the window again and then turns to follow him.

Their footsteps echo in the interior of the gloomy building.

Camera facing door through which they came in. The two walk toward it, De La Gardie leading, Wraxall following somewhat reluctantly. De La Gardie waits at the door for Wraxall. The two go out and the great door closes behind them.

Fade.

INT. LIBRARY, RABACK MANOR HOUSE.

A corner of the library at the Manor House. Wraxall is seated at a desk with masses of PAPER in front of him and big, LEATHER-BOUND VOLUMES piled high. The room is a different one to the study, dark, and with even more BOOKS. He turns over the pages, verifying references, occasionally pencilling notes.

C.S. Wraxall. He stops writing and passes his hand wearily across his forehead. He takes off his glasses and puts them on the desk in front of him.

Two-shot over Wraxall's shoulder, taking in De La Gardie, who sits reading in a corner.

DE LA GARDIE
(sympathetically)
Had enough for one day?

C.S. Wraxall.

WRAXALL
I have been going at it, rather. These old account books and journals are in such tiny writing. Tiring to the eyes. But it's fascinating material . . .

Two-shot. De La Gardie gets up and goes over to a DECANTER on a table.

DE LA GARDIE
You look as if a drink wouldn't come amiss. Whisky all right?

Wraxall smiling. He nods.

WRAXALL

Many thanks.

He pushes back his CHAIR from the desk as De La Gardie comes over and puts his DRINK down on the BLOTTER in front of him. De La Gardie goes back for his own drink and then takes a SEAT on the other side of the desk.

Side-shot taking in two men, the desk and the room behind them.

DE LA GARDIE

You've found something worth noting, then?

WRAXALL

There's always something.

(he waves his hand about the library)

And in the history of a family such as yours . . .

De La Gardie takes a sip from his GLASS with a smile.

DE LA GARDIE

No scandals, I hope?

WRAXALL

Oh, nothing like that, I assure you my dear Count! And in any event, I'm not looking for that sort of thing for such books as I write. I am thinking of the occasion in the sixteenth century when the peasantry attempted to burn this house.

DE LA GARDIE

Oh, that. Yes, I can see that it would interest. It's such an old story in our family, we tend to overlook such things.

WRAXALL

A stranger's eye . . .

(he consults a paper on the desk in front of him)

Take this, for instance. The writer, presumably the steward of the estate, refers to the suppression of the rising and the execution of the ringleaders. They had some barbaric customs in those times! Your ancestor, Count Magnus, seems to have been an extraordinary character.

C.S. De La Gardie. He is no longer smiling.

DE LA GARDIE

Count Magnus, yes.

His eyes flicker and he looks at something outside frame.

C.S. Wraxall. He follows the Count's gaze.

C.S. Dark PORTRAIT in oils of a singularly ugly man which hangs in a recess between the bookcases.

Two-shot. Wraxall looks across at the Count and puts down his glass.

WRAXALL

This is he, is it not?

DE LA GARDIE

Yes. And now, if you'll excuse me, Mr. Wraxall.

Wraxall gets up and the Count goes out by a door in the rear. Wraxall looks after him with a frown and then picks up his glass and goes over toward the portrait, the camera panning with him.

C.S. Portrait of Count Magnus.

C.S. Wraxall studies it closely.

C.U. The head only of the portrait. The eyes seem to gaze out at Wraxall.

M.S. Looking down on Wraxall as from the portrait.

C.U. The eyes of the portrait staring at Wraxall.

C.U. The upper half of Wraxall's face. He stares as if fascinated.

M.S. Taking in Wraxall and the portrait.

WRAXALL

(sotto voice)

So, Count Magnus, this is what you look like.

(he glances rather furtively round the room)

The De La Gardies don't seem very keen on you. I wonder why?

C.S. The portrait seems to sneer at Wraxall.

C.U. Wraxall. He gazes somewhat defiantly at the portrait.

WRAXALL

We shall see, Count Magnus. I intend to find out a little more about you before I leave Raback.

M.S. Wraxall and portrait. He turns, startled, at a SUDDEN NOISE.

C.U. A LOG OF WOOD has fallen out of the BASKET into the FIREPLACE.

C.S. Wraxall's startled expression relaxes. He hurries forward.

L.S. Wraxall goes forward to the fireplace and replaces the log with the TONGS. Then he glances around the room, dusts his hands and goes out the far door.

C.S. The portrait. Count Magnus appears to be following Wraxall with his eyes.

Fade.

EXT. RABACK CHURCH AND MAUSOLEUM.

L.S. SILHOUETTES, the plateau and church buildings. Tiny figure of Wraxall is walking along toward them.

Wraxall is approaching the church.

L.S. A tall, dark figure is silhouette walks rapidly away from the far side of the church.

M.S. Wraxall. He looks puzzled, half-raises his HAT in salute.

WRAXALL

That reminds me, I must ask the pastor for the keys.

He stares off into the distance.

L.S. The silhouette of the tall figure strides away through the far trees.

L.S. Favouring the mausoleum buildings. Wraxall passes, a tiny, almost hesitant figure.

Fade.

EXT. THE INN AT RABACK.

M.S. The Inn Signboard.

Dissolve to:

INT. INN AT RAYBACK—PARLOUR.

A small private parlour at the inn. Wraxall sits drinking at a TABLE near the FIRE. Nielsen, the landlord, is pouring himself a DRINK, evidently at Wraxall's invitation. There is a SMALL BAR COUNTER to the right and BARRELS can be seen in rear of it. The landlord finishes drawing from a KEG into a PEWTER TANKARD and comes back to join Wraxall at the table.

WRAXALL

(raising his glass)

Your health, landlord.

NIELSEN

And in the same sentiments to you, sir.

At Wraxall's gesture he sits opposite him, near the fire.

WRAXALL

You may remember me, we were discussing the matter of the mausoleum at the church yesterday.

NIELSEN

(reluctantly)

Oh, that.

(he looks round as though they might be overheard, and takes up his PIPE.)

There are a great many people in these parts who would rather forget that. I must apologise for my behaviour yesterday. My wife is rather particular on the subject.

Two-shot, closer and from a different angle. Wraxall draws his CHAIR closer to the table. He looks interested.

WRAXALL

I quite understand. I look at these matters from a different viewpoint. From that of the impartial historian, you understand. And the mausoleum and the history of Count Magnus in particular, are part of that research.

NIELSEN

(he draws on his pipe sceptically)

That's as may be, Mr. Wraxall. But if you knew as much about the history of that gentleman and the folk hereabout, mayhap you wouldn't be so curious.

C.U. Wraxall.

WRAXALL

You know something else, then? There wasn't much about him in the records I examined . . .

C.U. Nielsen.

NIELSEN

(he puts his finger to his nose and draws closer)

That's understandable. The Count wouldn't let those papers be seen by strangers. They're not over fond of him in the family. He was a black-hearted character, by all accounts.

Two-shot. Wraxall and Nielsen at table.

WRAXALL

You mean that not all the records have been put at my disposal?

NIELSEN

(a little embarrassed)

That's really not for me to say, sir. But from what I understand, the De La Gardies aren't anxious to discuss him or let the records relating to his life be seen by others. And I can't say that I blame them.

C.S. Wraxall. He looks intently at Nielsen.

WRAXALL

What exactly do you know?

C.S. Nielsen. He looks round him again and drains his tankard.

NIELSEN

By rights, Mr. Wraxall, I shouldn't say anything about this at all. But it wouldn't be fair to you not to—as a warning, as it were.

M.S. The two men.

WRAXALL

A warning? How do you mean?

(he glances down at Neilsen's tankard)

Have another. And you might draw me a second glass as well. This is an uncommonly fine Madeira.

NIELSEN

Certainly, sir. And thank you.

He gets up and goes over to the bar, the camera panning with him.

C.S. Wraxall looks after him for a moment. He looks thoughtful, but controls his patience until Nielsen comes back into the frame and puts down the glass.

L.S. Wraxall at table, the landlord by the bar. He draws himself another tankard and then comes back to take his seat. The two men drink for a moment or two in silence, the only sound the CRACKLING of the fire.

NIELSEN (CONT'D)

(he looks fixedly at Wraxall)

A warning, sir, I said, and a warning I meant. Count Magnus' is not a history to be delved into. He's buried yonder in the mausoleum. They keep it locked.

WRAXALL

But for what purpose?

NIELSEN

Good reason enough, sir. You wouldn't ask if you knew the Count's history.

(he lowers his voice and leans forward)

Take my advice, Mr. Wraxall, and leave Count Magnus out of your book.

C.S. Wraxall. He smiles thinly.

WRAXALL

But you can't leave it there, Mr. Nielsen. I must know more.

What is it about Count Magnus De La Gardie that must be kept hidden? The mausoleum cannot be that private. I saw the sexton coming out only yesterday.

C.S. Nielsen. He looks startled. He moistens his lips and appears to be about to blurt out something. Then he stops himself and speaks more deliberately.

NIELSEN

The sexton's away.

Two-shot.

WRAXALL

The Pastor, then. I saw him again only this evening coming from the door.

NIELSEN

(shakes his head)

The Pastor never goes there. It must have been some traveller like yourself, who failed to gain entrance.

WRAXALL

This is probably the answer. But I really should like to know more about Count Magnus.

C.S. Nielsen.

NIESEN

Very well, sir. I think I owe it to my guest to be outspoken. Let me just say this: that the Count was a monster of cruelty, who set the countryside round about ablaze with hatred in his time. He was a sadist and a torturer and he ruled with a rod of iron.

Two-shot.

WRAXALL

(his face clears)

I see! No wonder the De La Gardies wouldn't be keen to let people see these records. I understand now.

NIELSEN

Not quite, sir. As I said, the Count was a sadist and a torturer. He had whole families burned in their homes. And some said he had a compact with the Devil. I am not going from written records, you understand. These are stories which have been handed down from father to son, in these parts, for centuries.

WRAXALL

(with a slight smile)

Which may account for their distortion.

Two-shot. Change of angle.

NIELSEN

(lighting his pipe)

That's as may be, Mr. Wraxall, but the tales have been consistent over the years. There must be a great deal of truth in them.

WRAXALL

Do you believe in them?

C.S. Nielsen lights his pipe, puffing out great clouds of SMOKE. He speaks quietly and with great conviction.

NIELSEN

I do, sir. And I'd be easier in my mind if you believed them too.

Two-shot.

WRAXALL

What else do you know about Count Magnus?

NIELSEN

(shrugging)

If his tenants were late for work—they had to work for him certain days, as he was Lord of the Manor, you see—he had them flogged and branded in the manor yard. Or they were set upon the wooden horse there. I believe it's preserved in the building somewhere. And there were other things done in rooms deep within the Manor that no-one would speak about.

C.S. Wraxall over Nielsen's shoulder. The former shivers slightly.

WRAXALL

Yes, Mr. Nielsen, I can well see why the De La Gardies keep their ancestor's name from their lips. But what has that to do with the mausoleum?

Nielsen in reverse shot.

NIELSEN

(he listens for a moment as though expecting his wife to interrupt them)

Maybe nothing, Mr. Wraxall. But consider this. The Count went on the Black Pilgrimage.

Two-shot slightly angled to favour Wraxall.

WRAXALL

The Black Pilgrimage?

NIELSEN

(he nods)

The Black Pilgrimage, sir. And the locals say that he wasn't alone when he returned. That he brought something back with him.

A CLOCK suddenly starts CHIMING nine. Wraxall jumps and then looks sheepish. The landlord seems recalled to the present. He stands up abruptly and gathers his tankard.

Two-shot, wider angle, taking in room. The door opens suddenly. The landlord's wife appears. She waits silently as though implying the burden of all the work falls on her. Wraxall is about to ask more questions, but is obviously inhibited by the presence of Nielsen's wife.

NIELSEN (CONT'D)

I am called away to Skara early in the morning, Mr. Wraxall, so I may not see you until the evening. My wife and my staff will take care of all your wants.

WRAXALL

Thank you, Mr. Nielsen. You have been most kind.

Nielsen follows his wife out and the door closes behind them.

Closer shot. Wraxall and fire beyond. He stares at the door and then shivers suddenly, as though with cold. He goes over and sits down by the fire, holding out his hands to the blaze and rubbing them. He looks at the flames with a thoughtful expression on his face.

Fade.

INT. INN AT RABACK—WRAXALL'S ROOM.

Wraxall's bedroom in the inn. He is preparing for bed. He moves to and fro as he undresses, rather nervously, with his head on one side as though listening for something.

C.S. He winds his TRAVELLING CLOCK and puts it down by his bedside.

M.S. The camera tracks from a STRANGE CARVING in the bedroom toward the BED where Mr. Wraxall is settling himself.

M.S. Mr. Wraxall reaches out for his TYPESCRIPT MATERIAL on the BEDSIDE TABLE.

C.S. He starts to read through the material by the light of an OIL LAMP. He occasionally makes notes.

C.U. Mr. Wraxall's face as he reads to himself, his lips moving occasionally, as he repeats passages over. The name of De La Gardie and Count Magnus is heard.

NARRATOR (V.O.)

And yet Mr. Wraxall's curiosity about the De La Gardie family, and particularly Count Magnus, was greater than might be supposed. Why, he did not particularly know. He knew only that there was some mystery at the heart of the landlord's hinted comments. And Count De La Gardie himself had been somewhat enigmatic in his responses to his enquiries. The phrase, "the Black Pilgrimage" intrigued him and he returned to it again as he wrote his impressions that evening. And later, when he slept, such dreams as came to him, were far from pleasant ones . . .

EXT. WOODS OF COUNT MAGNUS' ESTATE.

C.U. Wraxall's face is beaded with sweat. The camera tracks back to show that he is in a STRANGE, MISTY LANDSCAPE back rounded by WOODLAND.

L.S. The camera pans across the misty landscape, then freezes as a HORRIFIC SCREAM is heard in the far distance.

C.U. Wraxall's faces registers shock. He turns, looking this way and that.

M.S. Camera tracks back from Wraxall as he starts to turn away. The DISTANT BAYING of dogs is heard.

L.S. Wraxall is running desperately through the thick undergrowth.

M.S. Camera tracks with Wraxall as he runs past the trees.

C.U. His anguished face.

L.S. Wraxall is toiling up a long slope. The sound of the dogs is louder now. The SHARP NOTE OF A HUNTING HORN is heard.

L.S. A distant knoll, wreathed in thin mist. The TALL DARK FIGURE of a man stands on the knoll. He is wearing a LONG CLOAK and a strange TRICORN HAT. He holds a HORN to his lips and blows.

C.S. Wraxall's anguished face as he flees through the trees. The baying of the hounds is much louder now.

M.S. Foliage rushing toward him as the camera plunges forwards.

C.U. FOLIAGE whips across Wraxall's face as he rushes on. The baying of the hounds still louder.

M.S. Wraxall staggers into a CLEARING. The camera moves back to emphasise his loneliness. Noise of hounds coming closer.

M.S. Wraxall rushes on again as sharp horn-notes sound. A noise to suggest the THUDDING OF HIS HEART on the sound-track.

C.S. GNARLED BOLE OF A TREE rushes at Wraxall.

C.U. Wraxall's despairing face as heart-beats, noise of hounds and hunting horn rise to a crescendo.

C.S. Foliage rushes at lens.

C.S. GNARLED BRANCHES rush at camera.

C.S. Wraxall screams. He falls out of frame as heart-beats, horns and dog-barks rise to a deafening climax.

L.S. Cloaked figure on knoll stands motionless. After a moment, the camera starts moving back.

Dissolve to:

INT. INN AT RABACK—WRAXALL'S BEDROOM.

M.S. Wraxall lying in bed. He is covered with perspiration, his eyes are wide open and he is sobbing with terror. Camera moves back from bed.

Fade.

INT. STUDY, RABACK MANOR HOUSE.

A door opens and Wraxall comes into the study. He is followed by De La Gardie. The camera moves with them down the study.

DE LA GARDIE

What are your plans for today?

WRAXALL

(smiling)

I intend to do a little quiet research among the account books.

They stop near the desk where Wraxall was doing his research the day before.

Two-shot, different angle. Wraxall sits down at the desk, while the Count remains standing.

WRAXALL (CONT'D)

I must say it will soothe my nerves after the night I spent...

DE LA GARDIE

(staring at him in a concerned manner)

You did not sleep well?

WRAXALL

Oh, it was nothing really. Just a curious dream . . .

M.S. With portrait of Count Magnus in the background, Count and Wraxall in foreground. Wraxall glances at the portrait.

DE LA GARDIE

(wryly)

Oh, well, if you will spend all your time among these old books . . .

(He sweeps his hand around to embrace the study.)

WRAXALL

(smiling at him)

It has its own rewards.

C.U. The Count looks at him enigmatically.

M.S. Wraxall sitting behind desk. He looks as though he is about to say something significant. He evidently thinks better of it. He is the first to lower his gaze.

M.S. The two of them with the room behind them

DE LA GARDIE

Would you like some coffee?

WRAXALL

(glancing at the CLOCK on the wall)

In about an hour, if I might. It will give me time to make a start.

DE LA GARDIE

You'll stay to lunch, of course.

(he moves forward to hand the other a PAPER)

M.S. The Count passes him the paper.

C.U. Wraxall.

WRAXALL

Thank you, yes.

Wraxall back to camera at desk in foreground. Count facing camera. De La Gardie smiles at Wraxall and goes out door in rear.

C.S. Portrait of Count Magnus.

C.S. The fire flickering in the grate.

M.S. Camera pans up from carpet to Wraxall at desk. He has BOOKS and BOX-FILES piled up. He is scribbling away at his notes.

M.S. Camera pans along shelves.

C.S. Latin title of ANCIENT BOOK.

C.S. Curious CARVING DETAIL on BOOKCASE.

C.S. A CHAINED VOLUME incredibly ancient and dilapidated. Wraxall's hand enters the frame and lifts it down.

M.S. Camera angled over Wraxall's shoulder as he reads the crabbed script of the book on the desk before him. The long chain goes off to the left towards the bookcase. He is muttering to himself as he translates.

C.S. Wraxall facing camera, book before him. He turns over the pages curiously and with mounting interest.

WRAXALL (CONT'D)

The Book of the Phoenix...

(grunts).

The Book of the Toad.

(very softly)

Most interesting...

C.S. Wraxall from overhead angle. He goes on turning and talking to himself.

WRAXALL (CONT'D)

The Book of Miriam. Turba Philosophorum.

He stops, his face trembling with excitement.

M.S. The camera moves in to show a blank page with some writing by Count Magnus himself, headed "Liber nigrae peregrinationis". Wraxall's hand enters the frame and his fingers trace out the ancient writing.

C.U. Wraxall's face. He licks his lips with excitement.

WRAXALL (CONT'D)

Count Magnus' own handwriting!

M.S. Wraxall gets up and consults an OLD VOLUME on a shelf and brings it over to the desk. A pause while he compares the handwriting. He puts the book back and returns to the desk, rubbing his hands.

WRAXALL (CONT'D)

Identical! Now perhaps we shall learn something further.

M.S. Different angle. Wraxall leans forward and picks up PEN and a WRITING PAD. He bends over the book and commences to translate.

NARRATOR (V.O.)

The papers on which Mr. Wraxall was just then engaged soon put his thoughts into another channel. They were of exceptional interest. Not only did they throw light upon the culture of that period in Sweden, but they were directly pertinent to Mr. Wraxall's current and indeed overwhelming interest. It is true, that the first volume indicated little of the material to come. They were account books in Count Magnus' own hand. But soon Mr. Wraxall had discovered a book of alchemical and other tracts dating from the sixteenth century. These related to very different matters indeed.

C.U. The crabbed sixteenth century script.

C.U. Mr. Wraxall's hand, writing the English translation.

C.U. Wraxall's face as he translates aloud with mounting interest.

WRAXALL

If any man desires... to obtain a long life... if he would obtain a faithful messenger... and see the blood of his enemies... it is necessary that he should first... go into the city of Chorazin... and there salute the prince...

Wraxall stops and looks puzzled.

C.U. The writing in the book. One word has been scored through. Wraxall's finger stops on it as his voice mumbles on, his words inaudible.

C.U. Wraxall's face.

WRAXALL (CONT'D)

(completing the sentence)

... the Prince of the Air.

M.S. Wraxall at desk. He turns the book round. He studies it.

C.U. Latin inscription in the margin, also in the Count's wtiting. "Quare reliqua hujus materiei inter secretiora."

WRAXALL (CONT'D)

(slowly, as he translates)

See the rest of this matter among the more private things.

C.U. Wraxall's absorbed face.

C.U. Portrait of Count Magnus.

M.S. Wraxall glances over at the portrait.

B.C.U. The eyes of Count Magnus stare at Wraxall out of the portrait.

C.S. Wraxall licks his lips and glances away.

C.S. Wraxall, camera over his shoulder, angled down at the book.

C.U. Wraxall's absorbed face, as he mumbles the words of the translation over to himself. A sharp noise makes him jump. He looks up, fear chasing across his features.

L.S. The door opens and a WOMAN SERVANT comes in with the COFFEE. De La Gardie following.

M.S. Wraxall at the desk. He smiles reassuringly, glances out of frame as though at the Count's portrait, and takes off his glasses. He reaches forward as the hand with the COFFEE POT comes into frame.

L.S. De La Gardie sits down at the other side of the desk, servant goes out and the couple start drinking the coffee as the camera draws back to the farthest limits of the room. Portrait of Count Magnus prominent to one side.

Fade.

EXT. RABACK CHURCH AND MAUSOLEUM.

L.S. Dusk, near the church. Wraxall is walking along, wrapped in thought.

M.S. Closer shot of Wraxall walking.

M.S. Tracking shot of the church.

C.S. Wraxall's abstracted face.

M.S. Mausoleum. Wraxall walks into the frame towards it.

M.S. Wraxall becomes aware of his surroundings. He looks towards the mausoleum.

M.S. Closer shot of mausoleum detail with great door.

M.S. Wraxall. He looks at the mausoleum and speaks aloud.

WRAXALL

Ah, Count Magnus, there you are.

A sudden WIND blows shadows across the façade of the mausoleum.

C.U. Wraxall's face, still looking toward the mausoleum.

WRAXALL (CONT'D)

I should dearly like to see you.

L.S. Wraxall in foreground with back to camera, church and mausoleum in background. A LOUD CLANGING NOISE is heard from the mausoleum interior.

C.U. Wraxall's face. He looks startled.

M.S. Wraxall hurries toward the church.

M.S. The great door. Wraxall hurries into frame and tries door of mausoleum.

C.S. Side view. The door RATTLES as Wraxall tries it, but it is evidently secured.

C.S. Wraxall's troubled face. The wind suddenly guts. He glances about him.

M.S. Wraxall puts his ear to the door panel and stands listening in a strained attitude.

WRAXALL (V.O.)

(interior monologue)

Like many solitary men I have the habit of speaking to myself aloud. I do not usually expect an answer. In this case, perhaps fortunately, there was no voice; only, I suppose, the woman who was cleaning the church dropped something on the floor whose clang startled me.

M.S. Wraxall moves away from the mausoleum door and glances around him hesitantly.

C.U. Wraxall's troubled face. The wind ruffles his hair.

WRAXALL (V.O.) (CONT'D)

Count Magnus, I think, sleeps soundly enough.

L.S. Wraxall with church behind him. He turns and resumes his walk.

EXT. THE INN AT RABACK.

M.S. Wraxall walks into frame and towards it.

M.S. The inn entrance. Wraxall, back to the camera goes into the inn.

INT. INN AT RABACK—ENTRANCE HALL.

Hall of the inn. DARK PANELLING, FLOWERS IN BOWLS, SEVERAL MIRRORS that relieve the gloom. Nielsen appears from a side parlour.

WRAXALL

Good evening, Mr. Nielsen. You had a successful trip to Skara?

NIELSEN

(smiling because his guest remembers his affairs)

Thank you, yes.

C.S. Hall. Wraxall is about to mount the STAIRS to his room, but Nielsen approaches and puts his hand on his arm.

NIELSEN

You were curious about the mausoleum at the church yesterday, Mr. Wraxall. I may be able to help you.

WRAXALL

(surprised)

How so?

Reverse shot, favouring Nielsen.

NIELSEN

The pastor is in the parlour yonder. I have been able to arrange a visit for you.

WRAXALL

That is extremely kind.

Two-shot. Hall.

NIELSEN

(who indicates a door off to the right, through which he has just emerged)

He is waiting to meet you.

He leads the way and opens the DOOR to the parlour. Wraxall hesitates a moment, looks around the hall and then follows.

INT. INN AT RABACK—PARLOUR.

Private parlour. A cheerful room, not previously seen; WHITE WALLS, SOME TAPESTRIES, a good fire in a BIG STONE FIREPLACE. The door at right opens and Wraxall and Nielsen enter, the camera panning left with them to embrace the whole room. A big DOG belonging to the landlord is lying on the hearth. PASTOR MOELLER is a plump, shortish man dressed in black with a suggestion of lace at the collar. He has a cheerful and garrulous manner and thick black hair going grey. He looks nothing like the figure Wraxall has twice seen at the mausoleum door. He sits in a WING CHAIR near the fire reading a LOCAL NEWSPAPER and with a GLASS OF WINE at his elbow. He gets up beaming as the two men enter. Nielsen goes forward to make the introductions.

NIELSEN

Pastor Moeller, this is Mr. Wraxall, the visitor I was telling you about.

C.S. Moeller, who puts out his hand.

MOELLER

I am charmed to meet someone who cares so much for the antiquities of our country.

C.U. Wraxall's face. He is staring at the Pastor, looking somewhat troubled. He forces a smile.

WRAXALL

I am sure our good friend Mr. Nielsen has been somewhat exaggerating my talents.

M.S. The group of three as Wraxall goes forward to shake hands with the Pastor. Nielsen is laughing at Wraxall's modesty. The pastor re-seats himself by the fire and Wraxall sits down opposite. A SERVING GIRL has come in and Nielsen orders two more glasses of wine, for himself and Wraxall.

C.S. Nielsen.

NIELSEN

Perhaps you gentlemen would prefer to be alone . . .? In which case I will leave you.

M.S. The group by the fire.

WRAXALL

By all means stay, Mr. Nielsen. There is nothing private. And you have been most helpful.

(he turns toward the Pastor)

M.S. The group, favouring Moeller.

MOELLER

(waiving Nielsen into a chair)

Do sit down, Mr. Nielsen. This concerns you as much as the rest of us.

M.S. The camera draws back from the table to embrace the entire room. Wraxall leans over and pats the dog. The only sound is the CRACKLING of the fire. The three men are silent until the girl brings the WINE and serves Nielsen and Wraxall. The camera pans with the girl to the door as she goes out.

M.S. The group by the fire. The two men silently toast Pastor Moeller, who lifts his own glass in salute.

MOELLER (CONT'D)

Mr. Nielsen tells me you are particularly interested in the church. I take it you have already seen Count De La Gardie? He has done an enormous amount of research on the matter.

WRAXALL

(nodding)

We are fellow enthusiasts!

The two men laugh politely, laughter in which Wraxall joins.

MOELLER

The Last Judgement is particularly fine.

WRAXALL

Yes, I have already had occasion to observe it.

NIELSEN

(sipping at his glass, and looking benevolently from one to the other)

We in Raback will be famous when Mr. Wraxall's book comes out.

Reverse shot, favouring Wraxall, who is obviously pleased.

WRAXALL

Oh, come now, Mr. Nielsen. Hardly that!

MOELLER

You under-estimate your own worth, sir. I have already perused your book on Brittany. The Count is quite an enthusiast of yours. And I must say the book revealed considerable erudition and scholarship.

WRAXALL

You are too kind. It does seem as though Count De La Gardie is my champion in these parts.

(he picks up his wineglass and drinks to cover his embarrassment)

C.S. Favouring Nielsen. The landlord looks keenly at both of them and puts down his glass.

NIELSEN

The Pastor was saying that tomorrow would be convenient for visiting the mausoleum.

M.S. The group of three.

MOELLER

Yes, yes, I was quite forgetting. Shall we say tomorrow at three o'clock? I will be there with the keys.

WRAXALL

Excellent.

MOELLER

That is settled, then.

M.S. Wraxall settles back in his chair and fondles the stem of his wineglass.

WRAXALL

My Bible scholarship is a little rusty. Can you enlighten me about something?

M.S. Group, favouring Moeller.

MOELLER

Certainly, if my own scholarship is not too rusty, also.

The three men laugh.

WRAXALL

Can you tell me anything about Chorazin?

C.U. Moeller. The Pastor looks a little disconcerted. He looks swiftly out of frame toward Nielsen.

C.U. Nielsen. He shoots a glance of warning at Moeller.

C.U. Wraxall. He waits patiently, glancing from one to the other.

C.U. Moeller. He licks his lips and tries to appear at ease.

MOELLER

Well, you surely remember that that place was once denounced for blasphemy?

M.S. The group.

WRAXALL

To be sure. It is, I suppose, quite a ruin now?

MOELLER

So I expect. I have heard some of our old priests say that Anti-Christ is to be born there. And there are tales . . .

C.U. Wraxall. His face is alive with interest.

WRAXALL

Ah! What tales are these?

M.S. The group. Moeller gets up abruptly and looks significantly at the CLOCK on the wall. He smiles thinly.

MOELLER

Tales, I was going to say my dear Herr Wraxall, which I have quite forgotten. And now, if you will forgive me, I have to prepare a sermon.

Nielsen and Wraxall get up too, Wraxall looking rather disappointed.

M.S. The group from different angle, with door in background. Moeller comes forward to Wraxall to shake hands.

MOELLER (CONT'D)

Goodnight. Until tomorrow, then.

WRAXALL

Goodnight.

Nielsen goes forward with the Pastor and the door closes behind them to a mumble of conversation. Wraxall looks after them and then crosses the room, deep in thought.

M.S. Favouring the fireplace. Nielsen's half-empty glass is still on the table. Wraxall sinks back into his chair and takes another sip from his glass.

L.S. Reverse angle. Wraxall by fire with door in background. Nielsen comes back, closes door and then comes over to the fire and pokes it in C.U. Flames flicker up past the camera lens. He goes over to his chair and resumes his seat.

M.S. Normal angle, with the two men. Wraxall looks across at Nielsen.

WRAXALL (CONT'D)

Herr Nielsen, I have found out something about the Black Pilgrimage. You may as well tell me what you know.

C.U. Nielsen in the act of drinking. He looks startled and puts down his glass. He stares at Wraxall without speaking.

M.S. The two men, favouring Wraxall.

WRAXALL (CONT'D)

What did the Count bring back with him?

C.U. Dog raises its head to look at the two men in the firelight.

C.S. Nielsen. He glances round as though they may be overheard and then draws his chair closer to his companion.

NIELSEN

Mr. Wraxall, I can tell you this one little tale. But nothing more. No more, you understand?

M.S. Reverse shot, favouring Wraxall. He nods without speaking.

M.S. Wraxall in foreground, camera facing Nielsen. Track in to hold Nielsen's face in close-up as he starts to speak.

NIELSEN (CONT'D)

In my grandfather's time—which is almost a hundred years ago, of course—there were two men in these parts who hated Count Magnus' memory. They said, "This Count has been dead a very long time. We do not care for him. We will go tonight and have a free hunt in his wood".

C.S. Wraxall. He nods.

WRAXALL

Yes, I know it. That would be the long wood on the hill behind Raback?

M.S. favouring Nielsen. He picks up his glass again before replying.

NIELSEN

Yes. Well, there were those in this very inn who advised them not to go. They said there are persons walking who should be resting. They took no notice . . .

Dissolve to (Landlord's voice continues over new scenes):

INT. INN AT RABACK—PARLOUR—FLASHBACK

L.S. The Inn parlour which is full of PEOPLE dressed in the fashion of a HUNDRED YEARS EARLIER i.e. about 1820. The two men, HANS THORBJORN and ANDERS BJORNSEN are both young and strong. They are dressed in rough country fashion, with heavy STAFFS and Game Bags held by straps over their shoulders. There is much laughter from them as they drink from TANKARDS and shake their heads. They seem much amused at the credulity of the village people.

C.U. Thorbjorn.

C.U. Bjornsen.

C.U. of worried face of VILLAGE ANCIENT who is holding forth.

M.S. The two men seen over the heads of the cluster of villagers who surround them.

LANDLORD (V.O.)

They could not be dissuaded. They were skilled in woodcraft. There were no forest-men to keep the wood because no-one wished to hunt there. The family were not at the Manor House. These men could do as they wished. They were determined to go to the wood that night.

M.S. The two men shake their heads, still laughing. They put down their pewter tankards. They push their way through the crowd.

C.S. NIELSEN'S GRANDFATHER, a fine-looking man with a close trimmed beard looks after them with a worried expression.

M.S. The two men, pleased with the impression they have created, pause at the DOOR, smiling.

M.S. The villagers. A WOMAN looks at her COMPANION and shrugs helplessly.

C.S. A YOUNG SERVING-GIRL, evidently attached to one of the young men. She looks frightened and bites her lip as she glances appealingly toward the door.

M.S. The two men wave and disappear through the door.

C.S. Nielsen stands looking after them.

Dissolve to:

EXT. INN AT RABACK—BALCONY AT INN—FLASHBACK.

M.S. A wrought-iron balcony which leads from one of the second floor rooms at the rear of the inn. Nielsen is sitting on the balcony looking across to the WOOD in the distance. With him are THREE OF THE VILLAGERS who have already been glimpsed in the earlier scene. They have a TABLE in front of them and are smoking and drinking to pass the time. It is a fine, moonlight night.

EXT. WOODS OF COUNT MAGNUS' ESTATE—FLASHBACK.

L.S. The sombre wood bathed in MOONLIGHT. NIGHT NOISES, CRICKETS etc. A LIGHT WIND stirs the branches. The wood is seen from far off.

C.S. Branches stir uneasily in the wind.

C.S. Long shadows ripple across the ground.

L.S. The two men, backs to camera, walk down a LONG GLADE away from the camera. They glance fearfully around them as they go and clasp their staves firmly.

L.S. The camera pans across a long, forbidding stretch of woodland, the moonlight throwing distorted shadows of the gnarled and twisted trees. A branch CRACKS and the camera stops its movement momentarily, before passing on.

C.U. Moonlight trembling on the surface of DARK WATER in one of the forest glades.

M.S. A tall, thin SHADOW slips across the ground.

EXT. INN AT RABACK—BALCONY AT INN—FLASHBACK.

The balcony. The group, Nielsen in the middle, are staring out of frame in tense expectancy. One of the men is about to say something, but Nielsen puts out his hand to stop him.

EXT. WOODS OF COUNT MAGNUS' ESTATE—FLASHBACK.

L.S. The wood and the distant landscape, bathed in moonlight. Everything looks incredibly peaceful. Suddenly, from far away there is a long and VERY TERRIBLE SCREAM.

EXT. INN AT RABACK—BALCONY AT INN—FLASHBACK.

Nielsen starts to his feet in terror. The others look at him, horror-stricken. Then Nielsen resumes his seat and picks up his DRINK with a shaky hand.

C.U. Nielsen. His lips are trembling. He finishes his drink and puts it down.

EXT. WOODS OF COUNT MAGNUS' ESTATE—FLASHBACK.

L.S. The wood and the landscape. There is no other sound and all the animal noises have stopped.

EXT. INN AT RABACK—BALCONY AT INN—FLASHBACK.

M.S. The group huddled fearfully at the table.

C.S. The camera pans slowly over the four fearful faces coming to rest on a GROTESQUE CARVING, part of a CEILING POST of the balcony.

LANDLORD (V.O.)

And to the wood they went. My grandfather was sitting on a balcony at the rear of the inn here. It was summer-time and a bright, moonlight night. From the balcony he could see out to the wood and the far hills and from that vantage point he could hear extremely well, because sound carries in that natural basin. He had a few close friends in with him for company and they were drinking and smoking while they pondered on the foolhardiness of young people who cannot resist meddling in things which do not concern them. At first they heard nothing at all; the time wore on. It was long after midnight, but still they sat there, half apprehensive, not really knowing what they expected to happen. But then something did happen. Something terrible and inexplicable. It was just after one o'clock when they heard someone scream, just as if the most inside part of his soul was twisted out of him. All of them on the balcony caught hold of each other and they sat so for three-quarters of an hour.

M.S. The balcony. The camera pans up from railing to reveal them sitting at the table, in roughly the same positions, but looking drawn and exhausted. The bottles on the table are almost empty. It is evident that they have been there for several hours.

C.S. Nielsen's haggard face as he gazes out of frame.

EXT. WOODS OF COUNT MAGNUS' ESTATE—FLASHBACK.

L.S. The camera pans slowly across the woodland from the vantage point of the balcony.

L.S. Camera angled down a LONG GROVE OF TREES a little nearer the house. A terrible, echoing LAUGH sounds.

EXT. INN AT RABACK—BALCONY AT INN—FLASHBACK.

M.S. The balcony. The men start in terror.

C.U. Nielsen's trembling face.

C.U. One of the men drops his PIPE from his mouth.

C.U. Terrified face of third man. He turns to . . .

C.U. . . . equally frightened face of fourth man.

EXT. WOODS OF COUNT MAGNUS' ESTATE—FLASHBACK.

M.S. Long shadows of trees dance grotesquely over the ground.

EXT. INN AT RABACK—BALCONY AT INN—FLASHBACK.

M.S. The landlord comes forward to the edge of the balcony and peers forward. The shadows dance over his face too.

M.S. The group on the balcony. Nielsen starts as a GREAT CLANGING NOISE sounds. The sound continues over . . .

EXT. WOODS OF COUNT MAGNUS' ESTATE—FLASHBACK.

L.S. The dark grove of trees. The clanging noise still echoes through the glade.

EXT. INN AT RABACK—BALCONY AT INN—FLASHBACK.

M.S. The group. Nielsen crosses himself and returns quickly to the table. The group sit there as the scene

Fade.

LANDLORD (V.O.) (CONT'D)

The night passed with interminable slowness. They must have sat there for hours, listening to the wind creaking in the branches, but there was no further sound from the men of their village who had gone into Count Magnus' wood. But presently, they heard something else, only about three hundred ells off. It may sound nothing to you sitting here in comfort. But it was the most dreadful sound in the world to my grandfather and the men who were with him that evening. Someone came to the edge of the wood and laughed out loud. It was not one of those two men who had walked so confidently out of the inn, who laughed and indeed, they have all of them said it was not any man at all. And after that they heard a great door shut . . .

EXT. COUNTRYSIDE NEARBY—FLASHBACK.

Daytime. Nielsen and the three other men walking up the PATH to a doorway in the PRIEST'S HOUSE.

M.S. Door opens and PRIEST comes out to speak with them. He looks white.

M.S. Priest looks at Nielsen and reaches back inside door for his GOWN and RUFF.

EXT. WOODS OF COUNT MAGNUS' ESTATE—FLASHBACK.

M.S. The FOUR MEN led by the priest are walking up a path near the FOREST. Two of the men carry a BIER.

M.S. Tracking shot of the group between the trees.

C.U. Priest's face. He is trembling and mumbling a prayer under his breath.

C.U. Nielsen's face. He looks fearfully around him.

M.S. The other three men walking in group.

M.S. A deeper part of the forest. The men are coming along in far distance toward the camera.

M.S. Tracking shot of gloomy trees gliding past frame. The sun does not penetrate much in here, and there is much shadow which throws distorted patterns.

M.S. Shooting towards group coming toward camera. Priest stops and starts. Group huddle behind him. Priest looks round, gathers his courage and comes toward camera. HANDS come out from behind camera into foreground of frame and make pushing motions. Priest comes forward, horror on his face, and takes hold of hands.

C.S. Face of Anders Bjornsen. His eyes are wide open and staring, but he sees nothing. He does nothing but mumble and push something away with his hands.

M.S. The group of men surround Bjornsen. They all look terrified.

M.S. Two of the men lead Bjornsen gently away. He is still moaning and making pushing motions.

M.S. A THICKET a little further off. Low angle shot with branches across frame. Priest and Nielsen, seen through branches, approach and stop near camera, staring downward.

C.U. Priest's face. He is trembling. He crosses himself.

C.U. Nielsen's horrified face as he stares downward.

M.S. Priest and Nielsen in foreground, two men leading Bjornsen going away down path in background. Fourth man with bier in between two groups. He comes up curiously, but Nielsen turns and stops him. He goes back to fetch a BLANKET from the bier. He comes toward camera, watched by priest. He carefully puts the blanket over the camera lens, blotting out the scene.

L.S. Nielsen and other men are carrying the bier back through the forest in the far distance. The priest walks behind. He is carrying a PRAYER BOOK and apparently reading something from it. A long way in front the two other men are leading Bjornsen back to the village.

M.S. The group with the bier. The leading man stumbles and the bier tips.

C.U. Nielsen's startled face.

M.S. The CORPSE falls tumbled in the blanket.

M.S. The priest stops in mid-stride, startled.

M.S. The three men looking down at the corpse.

B.C.U. The eyes of the dead man staring up at them; the rest of the face is hidden by the edge of the blanket.

M.S. The priest turns to the villager and speaks to him. The villager walks away.

Extreme L.S. The priest and Nielsen stand away from the blanketed corpse on the ground. The villager a long way off along the path.

M.S. WIND ruffles the branches.

M.S. Villager walks back along the path. He carries a SPADE.

M.S. Nielsen and priest in foreground. Villager walks up to them.

M.S. Villager starts digging grave, with priest and Nielsen in background.

Extreme L.S. The group. The grave is now dug. Nielsen and the priest start lifting the body toward it.

Fade.

LANDLORD (V.O.) (CONT'D)

Then, when it became light they went to the priest. They asked him to put on his gown and ruff and come to bury the men, Anders Bjornsen and Hans Thorbjorn. You see, they were certain that the two were dead. So they went to the wood. My grandfather never forgot this. He said they were all like dead men themselves. The priest, too, was in a white fear. He told them he had heard a cry in the night and someone laugh afterwards. And he could not forget it; he would never be able to sleep again. They found the two men in the wood. Hans Thorbjorn was standing with his back against a tree and all the time he was trying to push something away from him with his hands—something that wasn't there. So he was not dead. They led him away and took him to his house at Nykjoping and he died before the winter; but he went on pushing with his hands. Also Anders Bjornsen was there; but he was dead. And I tell you this about Anders Bjornsen; he was once a beautiful man but now his face was not there, because the flesh of it was all sucked away off the bones. You understand that? And they laid a blanket over his face and laid him on a bier they had brought and started back to the village, the priest reading a prayer for the dead. Then the one who was carrying the head of the bier stumbled and the corpse tumbled on to the ground. And when they looked down they saw the blanket had fallen away and the eyes of Anders Bjornsen were staring up at them because there was nothing to close over them. And they could not bear that. So the priest sent for a spade and they buried him where he fell.

INT. INN AT RABACK—PARLOUR.

M.S. The inn parlour with Nielsen and Wraxall. The fire has burned low. Wraxall's wine-glass is empty. He shivers suddenly as the landlord finishes talking. The dog on the hearth whines in its sleep.

C.U. Nielsen's expression is grim.

C.U. Wraxall looks shaken. There is nothing to say.

L.S. embracing entire room. Nielsen gets up suddenly and drains his glass. He goes over to the fireplace and starts putting more LOGS on the fire. Wraxall is sitting in a very pensive mood.

Fade.

EXT. RABACK CHURCH AND MAUSOLEUM.

L.S. A white road with trees in background. A PONY AND TRAP is bowling briskly along toward the camera. As it comes closer we see it contains Wraxall and De La Gardie, with Pastor Moeller driving.

C.U. Wraxall.He is smiling.

C.U. De La Gardie's face is thoughtful.

C.U. Moeller. In contrast to the others he has a worried look on his face.

M.S. Taking in all three occupants of the trap.

MOELLER

You understand, there is not a great deal to see in the mausoleum.

WRAXALL

(cheerfully, looking at De La Gardie)

Oh, yes. But it will make my book more complete. And I

may as well have a look while I am here.

C.U. De La Gardie.

DE LA GARDIE
At least it is one more thing that will remind you of Raback.

C.U. Wraxall.

WRAXALL
(glancing at the Count)
I shall not need reminding, I can assure you.

C.U. De La Gardie looks enigmatically towards Moeller.

M.S. Taking in all three in the trap. Moeller half-glances around and smiles quietly to himself.

L.S. Church buildings to left of frame. Trap approaches from distance and comes to stop in front of church, in M.S.

M.S. The trap and occupants. Moeller descends and ties the horse to a white post at the side of the road. Wraxall and the Count get down.

M.S. Moeller in foreground.

MOELLER
I won't be a moment. I just have to fetch the key.

DE LA GARDIE
I will stay here, if you don't mind.

Wraxall nods. He then follows Moeller. The Count waits by the trap. He avoids looking at the mausoleum, seen in the background, and faces the church.

M.S. The CHURCH DOOR. Moeller opens it. It is unlocked. He half-turns and smiles as Wraxall joins him.

INT. RABACK CHURCH.

M.S. Facing door of church. The two men come in as the camera tracks back with them. Moeller stands for a moment in silence, as though he has forgotten his errand. The camera pulls back, holding the two men in the silence of the vast interior.

M.S. Down from gallery, with two men.

M.S. Two men from floor level.

MOELLER

(with a smile)

I have so little call for the mausoleum key I hardly remember where it is kept myself.

WRAXALL

I am sorry to trouble you in the matter.

C.S. Moeller. He remembers and starts moving toward the PULPIT.

MOELLER

It is no trouble, I assure you.

C.S. A dusty detail of the CARVING under the pulpit. The camera pulls back as Moeller comes into frame, his back to camera. He pulls the HEAD of one of the TINY FIGURES. The head forms the knob of a MINUTE CUPBOARD, which opens outward.

M.S. Wraxall moves forward to get a better look at what the Pastor is doing.

C.U. Moeller's hand takes an enormous IRON KEY out of the cupboard interior.

C.U. Wraxall's intent face. He knows where the mausoleum key is kept now.

M.S. The two men, with the pulpit behind them. Wraxall moves casually away and glances round the church as Moeller closes the cupboard and re-joins him.

C.S. Moeller.

MOELLER (CONT'D)

(satisfaction on his face)

I have remembered, you see!

M.S. Wraxall in foreground. He chuckles as they move away.

L.S. The two men walk away from camera, back towards the great church door. It opens as they go out. Light floods in momentarily and then there is a great slamming noise, reminiscent of the moment in the Nielsen grandfather's narrative. It echoes around the church. Hold for a moment longer.

EXT. RABACK CHURCH AND MAUSOLEUM.

M.S. De La Gardie stands by the trap. He is patting the horse's head. The two men walk toward him.

M.S. From the opposite angle, past the Count and the horse. Moeller and Wraxall come toward him. Moeller holds the key and wags it toward him.

MOELLER (CONT'D)

There you are, you see!

C.U. De La Gardie. He laughs. He turns toward Wraxall.

M.S. Three-shot.

DE LA GARDIE

(to Wraxall)

I remember once when Pastor Moeller took three days to find the key! The visit from the Church Commission . . .

(he turns to Moeller, still smiling)

M.S. Favouring Moeller. He smiles too, wagging the key toward the Count playfully.

MOELLER

Ah, my dear Count, there was some excuse that time. We kept the key in the vestry in those days . . .

M.S. Favouring Wraxall. The two men start to turn away, the Count remaining where he is.

WRAXALL

Will you not come in with us?

M.S. Towards Count. He shivers and looks round him uneasily.

DE LA GARDIE

I think not, Mr. Wraxall. I will wait in the trap if you don't mind.

L.S. Count in foreground. Wraxall comes toward De La Gardie. Moeller is seen in background unlocking the mausoleum door.

M.S. Wraxall and the Count.

WRAXALL

I shan't be long.

C.U. De La Gardie's face. He nods, his eyes sombre.

C.U. Wraxall. He smiles reassuringly.

L.S. Wraxall moves away from Count and goes toward Pastor, who is still struggling with the key.

M.S. The Count gets up into the trap and seats himself.

M.S. THE PORCH. The Pastor is mumbling to himself as he exerts pressure on the key. He turns as Wraxall comes into frame.

MOELLER

I'm afraid it's a long time since this door has been opened.

WRAXALL

Oh, surely not. I saw someone coming out just the other day. In fact I thought it was you.

C.U. Moeller's face. It expresses alarm. He looks at Wraxall sharply.

M.S. The two men at the door.

MOELLER

I can assure you that you must be mistaken. To my knowledge this door has not been opened for more than a year.

C.U. Wraxall. His face expresses surprise. He is about to say something, but thinks better of it.

M.S. Favouring Moeller. He assumes a less serious expression.

MOELLER (CONT'D)

This is a small place, Mr. Wraxall. We get so few scholars. The average visitor is bored with the church. The mausoleum would be too much altogether.

(Moeller smiles and after a moment Wraxall smiles too.)

C.U. The key in the door and Moeller's hand trying to force it round.

M.S. The two men. Wraxall glances round.

L.S. The Count sitting in the trap, a silhouette against the cloudy sky.

M.S. The two men by the door. Wraxall hurries forward and puts his hand on the key. The two men struggle silently for a moment to force it round.

C.U. Moeller's concentrated face.

C.U. Wraxall He sets his teeth.

C.U. The giant key in the lock with the two men's tensed hands. It suddenly gives with a SCREAMING noise.

INT. MAUSOLEUM.

M.S. Inside the mausoleum, looking toward the door. A shadowy, echoing place. The SCREAM of the key in the lock still ECHOES round the interior.

EXT. RABACK CHURCH AND MAUSOLEUM

M.S. The two men at the door. Moeller smiles at Wraxall.

MOELLER (CONT'D)
There you are. Not so difficult, you see. I must get this oiled.

M.S. Moeller takes the key out of the lock and opens the door. He goes into the mausoleum. Wraxall hesitates and looks back.

L.S. De La Gardie in the trap. He waves.

M.S. Wraxall at the door. He waves back, smiles, and then follows the Pastor into the mausoleum.

INT. MAUSOLEUM.

A strange, octagonal building, much smaller than the church, but just as impressive. The two men approach from the door and walk toward the camera, Wraxall looking round curiously. Light filters down from the CUPOLA above; there are large marble 18th century MONUMENTS, with HERALDIC INSCRIPTIONS and much GILT.

M.S. The camera slides past the monuments.

C.U. Tracking shot, Moeller's face. The Pastor has seen all this many times and so is not impressed.

C.U. Tracking shot, Wraxall's enthusiastic face. He looks about him keenly.

L.S. From the dome area. Moeller and Wraxall walk into the shot far below. The central area of the mausoleum is occupied by three large COPPER SARCOPHAGI, covered with finely engraved ORNAMENTATION. The two men stop in the central area and stand as though lost in thought.

M.S. The two men from ground level, standing looking at the tombs.

C.S. Wraxall as he gazes silently.

C.S. Shot of one of the sarcophagi. It has a large METAL CRUCIFIX on the lid.

C.U. Moeller's face.

MOELLER (CONT'D)

As you can see, these two have metal crucifixes on the lid. That is a common practice in Sweden, as you probably know.

C.U.Wraxall. He nods.

WRAXALL

This is remarkably interesting.

M.S. Two men and tombs.

MOELLER

I thought you'd find it so.

(he turns)

The tomb of Count Magnus is over there.

M.S. Two men, shooting past coffins, favouring Wraxall.

WRAXALL

Ah! That is really the object of my search.

(he comes forward a little)

M.S. Two men from different angle, favouring Moeller. Moeller smiles thinly.

MOELLER

I'll leave you to your researches, then. What shall I tell Count De La Gardie?

WRAXALL

(turning away from the caskets)

I shall probably be some time. Perhaps I should rejoin him at the Manor House. Unless he wishes to come in . . .?

M.S. Moeller. He shakes his head.

MOELLER

He will not do that.

(he looks around rather furtively)

M.S. Wraxall waiting, the monuments seen in the background. He is obviously eager to begin.

Two-shot, angled over Wraxall's shoulder to Moeller.

MOELLER (CONT'D)

I will fetch some oil from the vestry and attend to the lock. Would you please replace the key in the church when you have finished...

Two-shot. reverse angle, favouring Wraxall.

WRAXALL

Certainly. And please tell Count De La Gardie I will be no longer than necessary.

L.S. Showing the two men and the interior of the mausoleum. Moeller gives a slight bow and goes out, his feet raising ECHOES from the dome.

M.S. Wraxall. He looks up toward the dome. The camera pans up to come to rest on the cloudy light spilling in.

M.S. Wraxall. He looks about him for a moment, rubs his hands and then walks round the end of one of the tombs toward the sarcophagus of Count Magnus. The camera pans with him.

M.S. DETAIL of Count's tomb in left foreground as Wraxall comes toward it.

C.U. Wraxall's face. He examines the detail carefully.

M.S. The tomb of Count Magnus. It has a full-length EFFIGY of the Count engraved upon the lid. It should bear some resemblance to the tall figure already glimpsed upon the mound in Wraxall's dream.

M.S. Wraxall moves nearer. He adjusts his glasses and appears fascinated.

C.S. The FRIEZE of carved scenes round the edge of the sarcophagus. A BATTLE SCENE with CANNON belching smoke and PIKEMEN advancing.

C.U. Wraxall's intent face.

C.U. Camera pans slowly along edge of tomb, showing VARIOUS DETAILS. A MAN BEING HANGED; another being TORTURED.

M.S. Wraxall is scribbling notes in his book.

C.U. Detail of the torture scene.

C.U. Wraxall's face. There is a sudden sharp SOUND. A look of terror passes over Wraxall's features. He looks round.

L.S. Toward door. Moeller can be seen in the porch, fixing he lock. He waves nonchalantly toward Wraxall.

M.S. Wraxall looking toward the door. He returns the salute and bends toward the tomb again.

C.U. The TOMB PANEL. A MAN IS RUNNING at full speed among trees, his hair flying, his hands outstretched.

M.S. Wraxall taking notes.

C.U. Man running between trees. The camera pans to a strange, MONSTROUS FORM pursuing him. Short figure, in HOODED GARMENT which sweeps the ground. The hand outstretched is not a hand, but something like a TENTACLE. Brief flash only of this creature.

M.S. Wraxall looks somewhat startled. He moistens the tip of his PENCIL with his tongue and continues with his notes. He mumbles to himself as he writes.

WRAXALL

An allegorical representation of some kind . . . A fiend pursuing a hunted soul? The origin of the story of Count Magnus and his mysterious companion?

INT. M.S. Wraxall writing in front of the tomb.

INT. L.S. Moeller glimpsed at door. He finishes his oiling and walks away without glancing into the mausoleum.

INT. M.S. Wraxall, still writing. He scribbles and ends a sentence in his notebook with a flourish.

WRAXALL (CONT'D)

Let us just see how the huntsman is pictured. Doubtless a demon with a horn.

INT. C.U. TOMB DETAIL. Camera pans slowly from pursued man to a representation of a CLOAKED MAN standing on a hillock. He leans on a stick and watches the chase with interest.

INT. M.S. Wraxall crouching looking at the detail of the tomb. There is a sudden sharp NOISE in the distance and Wraxall straightens up quickly.

C.U. Wraxall's nervous face. He glances out of frame.

L.S. The church door. The Pastor has gone.

C.U. Wraxall turns back to the tomb. His glance is arrested.

C.U. A huge BRONZE PADLOCK securing the coffin lid.

M.S. Wraxall comes closer.

C.U. Another huge BRONZE PADLOCK.

M.S. Wraxall stands lost in thought. He puts a hand up to his face and strokes his chin.

C.U. The bonze padlock. The camera pans slowly along the lid to the second one. It pauses there and then continues its movement towards the third and last padlock. Only there is no padlock there, only the HASP through which it had been threaded.

M.S. Wraxall is puzzled. He looks about him and then down to the floor.

C.U. The padlock hasp. THE CAMERA PANS slowly down to stop on the THIRD PADLOCK, which is lying on the paving at the front of the tomb.

INT. C.U. Wraxall's face. His eyelids flicker with shock. He coughs nervously.

L.S. Wraxall is standing silently in front of the huge tomb looking at the padlock on the floor.

L.S. Looking from tomb toward the door. Wraxall stands hesitantly a moment, then puts his notebook away. He turns and walks back toward the door, his footsteps ECHOING eerily under the domed ceiling. He goes out, SLAMMING the door behind him. Follows the SOUND of the key being inserted in the lock. It SQUEAKS a little, but the Pastor's work has evidently been successful. The lock CLICKS home and the door RATTLES as Wraxall tries it. His FOOTSTEPS are heard going away outside. Slow pan down to padlock lying on paving in foreground.

Fade.

INT. STUDY, RABACK MANOR HOUSE.

M.S. Wraxall is seated at the desk talking animatedly with De La Gardie, who sits quite near to him. There are COFFEE CUPS on the desk in front of them, REFERENCE BOOKS, OLD VOLUMES. They are laughing as the scene fades in.

DE LA GARDIE

(smiling)

I hope the poor old gentleman got out of the church safely, in the end . . .?

WRAXALL

(adjusting his glasses with a reminiscent smile)

Oh, yes, the Dean made sure of that.

M.S. Angled to take in desk, the pair and the Count Magnus portrait in background. It is obviously late, the LAMPS lit, twilight outside, the time after dinner. Wraxall puts on his glasses and shoots a glance at his WATCH.

WRAXALL (CONT'D)

The time is getting on. But I would just like to finish the notes before leaving.

(he glances out the window and then toward the portrait)

DE LA GARDIE

By all means. Would you like me to leave you?

WRAXALL

(shaking his head)

Not unless you wish. I have grown to rely on your unique knowledge of these parts. In fact, I shall be sorry to leave.

C.S. De La Gardie looks at Wraxall intently.

DE LA GARDIE

And yet I think it would be best, Mr. Wraxall.

M.S. The men at the desk. Wraxall coughs awkwardly.

WRAXALL

I don't quite understand.

DE LA GARDIE

I can't say any more.

M.S. The pair look round rather guiltily as there is the sound of the door opening. The woman servant walks into the room.

M.S. De La Gardie, the half-open door seen behind him. He stands up.

DE LA GARDIE (CONT'D)

I hope you have everything you want?

(he smiles benevolently)

M.S. Over servant's shoulder to Wraxall. He shifts uneasily in his chair, coughing again. Servant starts clearing cups.

WRAXALL

Oh, certainly.

(he stands up, too)

M.S. Taking in all three. De La Gardie goes forward to shake hands with Wraxall.

DE LA GARDIE

I will say goodnight, then. I am just going to take a walk around the estate before it grows too dark.

M.S. Different angle, favouring Wraxall.

WRAXALL

Goodnight, then. Until tomorrow.

M.S. Looking toward De La Gardie and servant.

DE LA GARDIE

Until tomorrow.

L.S. Wraxall in foreground at desk. The other two walk away toward the door.

C.S. Wraxall looks after them, then glances out of frame at:

C.S. Count's Magnus' portrait.

M.S. Wraxall bows ironically toward it.

WRAXALL

Are you asleep or are you awake, Count Magnus? Because I must just visit you on my way back to take a note of the inscriptions.

C.S. The portrait of Count Magnus gazes ironically back.

M.S. Wraxall standing by the desk. He suddenly looks a little unsure of himself. He reaches for his notebook on the desk in front of him and tidies the PAPERS. Moeller and the Count pass the window at that moment, in animated conversation. The Count's laugh attracts Wraxall's attention. He turns and watches them until they are out of sight. Then he turns back to the desk, smiling. He is humming softly as he clears the desk for the night. Then he turns out the oil lamp. The room is now lit only by the flickering flames of the DYING FIRE. He walks to the door and closes it behind him as the camera is left looking at the SUNSET through the window.

Dissolve to:

EXT. RABACK CHURCH AND MAUSOLEUM.

M.S. Knoll near the church. Wraxall is walking rapidly toward it, evidently keen to reach it before the light fails. The camera tracks a little closer as Wraxall goes into the church.

M.S. Looking toward the church porch. Hold shot as trees wave in the wind, casting long shadows over ground. Wraxall comes out the porch with the key and walks toward mausoleum. Camera pans right with him to take in mausoleum buildings. WIND rises slightly.

L.S. Wraxall, a tiny figure at the church porch. Struggling with the key. The camera tracks slowly towards the mausoleum as Wraxall finally opens the door. He disappears into the interior as the camera still comes forward. Hold shot stationary for a while on the door. The shadows of the tree branches fall across the whiteness of the mausoleum.

INT. MAUSOLEUM.

L.S. Wraxall moves forward down the mausoleum, his feet raising ECHOES from the domed ceiling, the camera moving smoothly back with him. It is darker now and the interior is misty and shadowy.

M.S. Wraxall rummages in his KNAPSACK (which he sometimes carries, but not always. This evening, for the express purpose of the visit, he has brought it with him.) He brings out a powerful HAND-LANTERN and lights it.

M.S. The BEAM glances over the DETAIL of STATUARY and SARCOPHAGI.

C.U. The beam glances over the hanged man on the frieze round Count Magnus' tomb.

C.U. Detail the man fleeing from his enemy with the figure on the hill beyond.

M.S. Wraxall glimpsed behind the lantern beam as he comes forward.

C.U. Count Magnus effigy on top of the tomb. Face only. As the lantern-beam dances over it, it appears to be smiling.

M.S. Wraxall grunts to himself as he goes forward with the lantern. He props it on top of a nearby tomb to direct beam forward. He kneels down and starts making notes on a LATIN INSCRIPTION he finds there.

C.U. Wraxall's engrossed face. He mumbles to himself as he continues with the inscription. Then his eyes glance down and flicker with surprise.

C.S. Camera pans from Latin inscription down to floor and first padlock lying there. Pans and then passes to right to disclose second padlock lying there.

C.U. Wraxall looks nervously around him.

M.S. Wraxall with tomb. He bends down and picks up both padlocks.

WRAXALL

(softly, aloud)

I didn't realise there were two padlocks loose.

M.S. Wraxall by tomb, from different angle. He takes the two padlocks and puts them down on a LEDGE near a NARROW WINDOW, the camera moving with him and then coming back to the tomb and the lamplight again. It is appreciably darker in the mausoleum now and Wraxall gives an uneasy glance from right to left.

C.U. Count Magnus' face on tomb lid. With the approaching darkness and the difference in the shadows, it seems to wear a different expression now.

C.U. The edge of the tomb showing the two empty sets of padlock hasps for the lid.

C.U. Wraxall standing deep in thought. He moves his eyes from one hasp to the other.

C.U. The camera pans from the empty hasps to the one remaining padlock on the right.

C.U. Wraxall's absorbed face.

C.U. The remaining padlock.

M.S. Wraxall by coffin, lamplight on his face. He reaches forward almost dreamily and tests the third padlock with his hand.

C.U. Wraxall's hand trying to open the padlock.

C.U. Wraxall looks almost disappointed.

L.S. Wraxall alone in the silent mausoleum by the tomb. The light is dying out in the dome now.

C.U. Quick flash of hanged man on coffin frieze.

C.U. Battle scene on frieze.

C.U. Fleeing man.

C.U. Man on mound.

C.U. Count Magnus' coffin face.

(Jagged chords of music over images, or silence more effective?)

L.S. Wraxall stands by the coffin. His whispered words come over as commentary and seem to echo around the mausoleum.

WRAXALL

The remaining padlock stands firm. A spring lock, probably. I cannot open it. However, Count Magnus, had I succeeded in undoing it, I am almost afraid I should have taken the liberty of opening the sarcophagus and peeping at you. It's strange, the interest I feel in the personality of a grim old nobleman like you.

Closer shot of Wraxall by tomb. He relaxes, shrugs slightly and smiles to himself. He looks around the tomb-house and realises it is getting dark. He reaches to pick up his lamp.

L.S. Another angle, tomb in foreground. Wraxall goes away down the mausoleum, the wobbly lamplight outlining his figure from time to time.

C.U. The lamplight flashes over the sardonic mask of Count Magnus on the tomb. The ECHOING FOOTSTEPS of Wraxall continue and then the CLANGING REVERBERATION as the outer door is closed behind him.

Fade.

EXT. THE INN AT RAYBACK.

L.S. A bright, sunny morning, but windy. Wraxall has a springy step as he comes into frame, walking out the inn door.

EXT. RABACK MANOR HOUSE AND GROUNDS.

L.S. Path with Raback Manor House. Wraxall comes along with his knapsack in an almost jaunty manner.

L.S. The Manor House.

M.S. A path in front of the Manor. De La Gardie is coming down it. He waves jauntily at Wraxall.

M.S. Wraxall pauses and smiles. He waves back and then hurries forward.

L.S. Pathway along Manor House. Count in middle distance with the house behind him. Wraxall enters from behind camera and walks to meet him. The two meet midway between camera position and house.

M.S. Wraxall and the Count stand together talking. We are too far away to hear their conversation. Hold for a moment or two.

M.S. DUCKS on an ORNAMENTAL POND in the grounds are excitedly dipping their beaks into the water and swimming vigorously around. The bright sunlight throws clear-cut shadows of the tree branches across the sparkling water.

L.S. Angled across the pond. Wraxall and De La Gardie talking earnestly are walking slowly toward the entrance of the Manor House. They disappear inside it.

Dissolve to:

INT. LIBRARY, RABACK MANOR HOUSE.

L.S. Wraxall walks into frame, followed by De La Gardie. He is carrying Wraxall's knapsack and puts it down on the seat of a big, CARVED CHAIR. Sunlight spills in at the windows.

M.S. De La Gardie sits to right side of desk.

DE LA GARDIE

(smiling)

Surely not tombs and dusty books on a day like this, Mr. Wraxall?

C.S. Wraxall smiles.

WRAXALL

I am afraid the open air will have to wait. I still have much to do.

C.U. De La Gardie, shifts in his chair.

DE LA GARDIE

Even so, I should on no account overtax yourself, Mr. Wraxall. I could think of many more diverting pursuits on such a morning.

Two-shot, favouring Wraxall. De La Gardie is still sitting beside the desk.

WRAXALL

There is much in what you say, Count. But I am conscious of the inconvenience to which I must be putting you.

DE LA GARDIE

(smiling again)

Of course not, my dear sir. And now, if you will excuse me, I will leave you to continue with your labours.

He nods in a fatherly fashion and leaves the room, coming past the camera. The door is heard closing behind him. Wraxall stands somewhat awkwardly for a moment, before moving to sit down at the desk.

C.U. Wraxall looks abstractedly as the desk.

C.U. Details of JUMBLED BOOKS.

C.U. Scribbled detail of his NOTEBOOK.

C.U. Heavy, LEATHER-BOUND VOLUME in Latin.

C.U. Favouring Wraxall. He sighs heavily and looks out of frame.

M.S. Count Magnus portrait.

M.S. The Count re-enters silently. He looks long at Wraxall before he speaks.

DE LA GARDIE (CONT'D)

I meant well in what I said just now.

WRAXALL

(slightly embarrassed)

I am sure, my dear Count. I did not take your words amiss.

C.U. The Count.

DE LA GARDIE

You will be careful, won't you?

WRAXALL

(startled)

I am tired, it is true, and I haven't been sleeping very well lately.

(he pauses and then goes on, picking up courage)

You meant something else, didn't you?

C.U. De La Gardie. He nods.

DE LA GARDIE

Yes. It is not good for you to spend so much time in the mausoleum. You are over-sensitive and highly-strung. Such an atmosphere . . .

(he breaks off)

C.U. Wraxall. He looks at his host with understanding.

WRAXALL

I will be careful.

C.U. De La Gardie. He nods, without speaking, and then looks down at the desk.

L.S. Camera up high, as though from position of portrait, looking down toward Wraxall and Count at desk.

Dissolve to:

EXT. RABACK CHURCH AND MAUSOLEUM.

L.S. Susk. The church, with Wraxall silhouetted against skyline.

M.S. Wraxall with trees and skyline behind him. He looks towards the mausoleum.

WRAXALL

(aloud)

You see, I am still at Raback, Count Magnus. And I must confess your history engrosses me considerably . . .

Dissolve to:

M.S. Wraxall is opening the door of the mausoleum.

INT. MAUSOLEUM.

L.S. Wraxall is walking toward Count Magnus' tomb in the dusk of the interior.

M.S. Camera is tracking toward to tomb from Wraxall's viewpoint. His footsteps echo eerily in the dusk.

M.S. From reverse angle, corner of tomb in left-hand frame. Wraxall comes toward tomb and stops.

C.S. Wraxall's dreamy face as he looks at the tomb.

M.S. The tomb of Count Magnus. All is silent and still. Two empty padlock hasps prominent.

C.U. Wraxall.

WRAXALL

(as if to himself)

You were a great villain in your time, Count Magnus. But nevertheless I should still like to see you . . .

He breaks off at a strange NOISE.

C.S. Padlock bounces on to Wraxall's foot and on to floor.

M.S. Wraxall jumps back, startled.

C.U. Wraxall looks at tomb.

C.U. Empty padlock hasp.

C.U. Wraxall licks his lips. He looks downward.

C.U. The third padlock is lying on the floor.

Extreme C.U. Wraxall's eyes wide in fear.

M.S. Wraxall bends down to pick up the padlock.

C.U. His hand touches the padlock.

M.S. Wraxall stooped with the padlock in his hand. A harsh SCREAMING noise sounds.

C.U. Quick flash of Wraxall's startled face.

C.U. The COPPER HINGES on the other side of the tomb. They are moving.

C.U. Wraxall's appalled face, from above shooting down.

C.U. The lid of the sarcophagus is rising slowly up. Nothing but shadow inside.

M.S. Wraxall gives a low MOANING CRY. He straightens up with a supreme effort.

C.U. The lid comes up higher.

M.S. Camera rushes at Wraxall's appalled face.

M.S. Wraxall's trembling form is seen, with tomb cover in foreground. It rises up and blots out Wraxall's face.

M.S. Wraxall drops the padlock with a bubbling cry of horror. The hinges go on SCREAMING as he turns and rushes out of the church.

M.S. Reverse angle. Wraxall blunders into a heavy iron CANDELABRA STAND and dashes it to the ground.

EXT. RABACK CHURCH AND MAUSOLEUM.

M.S. Outside mausoleum. The door opens and Wraxall staggers out with a face like death.

L.S. The silhouette of his running figure on the skyline. He is running madly in the direction of the inn.

INT. MAUSOLEUM.

M.S. Near the door, shooting toward it. The door BANGS noisily in the rising wind, letting in alternate bars of light and shadow. Hold shot as wind HOWLS around the building.

Fade.

INT. INN AT RABACK—WRAXALL'S BEDROOM.

L.S. Mr. Wraxall's room at the inn. He opens the door and comes in hurriedly. He is pale and dishevelled, with trembling hands. He looks round the familiar room with relief. He crosses left of frame and washes his hands in a BASIN, pouring water from the bowl.

M.S. Wraxall completes his toilet.

C.S. He TOWELS his face. His eyes have a far-away look.

L.S. Wraxall picks up a COMB and starts to tidy his hair. Room and door seen behind him. Sudden KNOCK at door. Wraxall starts violently.

M.S. Wraxall is trembling. He gazes toward the door.

WRAXALL

(tremulously)

Come in.

L.S. Nielsen pokes his head round door cheerily.

NIELSEN

I thought I saw you come in, Mr. Wraxall. Dinner is in half an hour.

WRAXALL

Thank you.

NIELSEN

(hesitating)

Will your friends be dining with you?

Wraxall stares at him blankly

NIELSEN (CONT'D)

I mean the two men who came in with you.

M.S. Favouring Wraxall. He has a terrible face.

WRAXALL

(very deliberately)

There was nobody with me.

NIELSEN

(he looks puzzled)

As you say, Mr. Wraxall.

(he slowly closes the door)

L.S. Toward door. Wraxall in centre frame, BED in foreground. Wraxall stands looking toward door, then collapses on to bed.

Dissolve to:

M.S. The bedroom, bathed in moonlight. It is late at night. An owl HOOTS somewhere far away. The camera pans slowly to bed, with Wraxall lying in uneasy sleep. Hold and then move toward WINDOW, with tree-branches stirring in wind, throwing long shadows across floor. Camera moves slowly toward window to frame branches in moonlight.

L.S. The view toward Count Magnus' estate, indistinct in moonlight and with GROUND MIST. There is a furtive CRACKLING NOISE, as though something is moving through the undergrowth.

C.U. Wraxall's face. He moves uneasily in his sleep.

Dissolve to:

EXT. WOODS OF COUNT MAGNUS' ESTATE.

L.S. A high, wide place, bare with a few trees. Wraxall is fleeing for his life (slow motion?).

C.U. Wraxall's terrified face, hair streaming in wind.

M.S. Indistinct, hooded figure rushing at his heels.

L.S. The two fleeing figures on the wide plain, the camera pans up right from them to a hillock, where a tall, cloaked figure in a broad hat and leaning on a staff stands motionless.

M.S. Blurred trees are rushing past the camera. A terrifying SCREAM sounds and the camera movement suddenly stops. The picture begins to fade as life dies.

L.S. Very distant silhouette of the figure standing on the mound.

Dissolve to:

INT. INN AT RABACK—WRAXALL'S BEDROOM.

C.U. Wraxall in bed. SCREAM sounds again, as though it is in the room. Wraxall opens his eyes. His face is drenched with sweat. He gazes wildly about him, then buries his face in the PILLOW and sobs.

Fade.

EXT. LAKE WITH STEAMER.

L.S. The steamer landing. There is a bustle as about thirty or forty PASSENGERS wait to embark. MEN are loading the small vessel and the mountains are seen in the background. Wraxall and De La Gardie, dressed as though for travel, enter frame, along with other travellers.

C.S. The Count's animated face as he chats to Wraxall.

C.S. Wraxall. His manner is abstracted. He hardly hears the Count's words, but looks about him, among the passengers.

NARRATOR (V.O.)

The following morning Mr. Wraxall looked so pale that the Count suggested a steamer trip as a change from the long hours of study. In truth, Mr. Wraxall welcomed the suggestion. But he found, as soon as they had embarked upon the voyage, that he had exchanged one set of anxieties for another. There was something curious about his fellow passengers, he felt . . .

M.S. Wraxall and De La Gardie are being jostled up the gangway.

C.S. Wraxall looks sharply out of frame.

L.S. Tall man in a cloak and broad-brimmed hat and a short figure with some sort of hooded scarf disappear among the passengers embarking. Brief glimpse only.

EXT. ON BOARD STEAMER.

M.S. Wraxall and Count on deck of STEAMER, which is already under way. Wraxall looks about him worriedly.

Dissolve to:

L.S. Water boiling in steamer's wake; slow pan up to distant shore.

Wraxall and Count are seated on a BENCH; one or two OTHER PASSENGERS near. De La Gardie is doing most of the talking, Wraxall paying hardly any attention.

DE LA GARDIE

You see, isn't this delightful? You look much better.

WRAXALL

(turning to him with an effort)

Yes, delightful!

M.S. The Count smiles warmly.

C.U. Wraxall's face. He turns back to scan the passengers at the ship's side.

L.S. Near the bow, with water rushing by. The SHIP'S RAIL. A tall figure in a cloak and with a broad-brimmed hat is standing motionless looking across at the shore, back to camera. A short, broad figure with a hood is at his side. The figures remain motionless, but the wind whips their clothing out.

C.U. Wraxall stares, the wind whipping the hair into his eyes.

C.U. The Count, hair blowing in wind, smiles at him, oblivious of his agitation.

M.S. Upper half of cloaked figure. The head moves slightly, the broad-brimmed hat whipping in the wind.

M.S. Similar shot of short, hooded figure. No detail of faces or even back of head seen.

B.C.U. Wraxall's eyes widen in fear.

C.U. Count and Wraxall on bench. Wraxall stands up abruptly.

L.S. The two figures at the rail, clothing blowing in the wind.

C.S. Wraxall. He swallows and looks at De La Gardie.

C.S. De La Gardie looks at him puzzled.

M.S. De La Gardie stands up to join Wraxall. He starts to move forward, stops.

L.S. A group of PASSENGERS move across, momentarily obscuring the two figures at the rail.

L.S. Wraxall moves forward, the Count behind him.

M.S. The passengers move away. The frame is empty, nothing but the ship's rail, the sea and the distant shore are seen.

C.S. Wraxall looks about him wildly.

L.S. The empty rail with the sea foaming by.

M.S. A group of PASSENGERS pass the camera, revealing Wraxall and Count. Wraxall is looking wildly about him.

L.S. Two dark figures are glimpsed beyond other PASSENGERS up toward the bow.

L.S. Wraxall and De La Gardie are walking fast along the deck up toward the camera. The Count looks anxiously at Wraxall.

L.S. Passengers move away, revealing cloaked man and short hooded figure up near the rail at the bow.

M.S. Wraxall leaves De La Gardie and runs forward.

M.S. Camera tracks forward to the two figures at the rail.

M.S. Wraxall moves forward to the two figures at the rail.

C.U. Wraxall's face. It has on it the courage of desperation.

C.U. Wraxall's hand reaches out and touches the dark figure's shoulder.

C.U. Wraxall's face is sweating.

C.U. Cloaked figure turns. It is that of a PASTOR, with a severe WHITE COLLAR and a scholarly face.

M.S. Wraxall looks bewildered. De La Gardie comes rapidly up behind him and takes his arm.

M.S. The cloaked Pastor raises his hat, saying something in Swedish, which Wraxall doesn't understand. Camera pans right to hooded figure. It turns, revealing:

C.U. Face of an elderly NUN. She turns to Pastor and listens anxiously.

M.S. The entire group at the rails. Wraxall is embarrassed and stammering.

WRAXALL

I beg your pardon.

De La Gardie moves forward and says something in Swedish. The Pastor raises his hat again and answers him. The couple turn back to the rail as Wraxall and De La Gardie walk away.

M.S. Wraxall and Count walking toward camera. The Count looks at him anxiously. Wraxall looks over his shoulder with haunted eyes. He mops his forehead with a handkerchief.

Extreme L.S. of Pastor and nun at rail. There is subtle difference in their costume and figures now, from the figures Wraxall has glimpsed before.

M.S. The ship's wake boiling behind it.

Fade.

INT. INN AT RABACK—WRAXALL'S BEDROOM.

L.S. Wraxall is writing in his room. He gazes fearfully around him from time to time.

M.S. Over his shoulder. He is writing a LETTER. Camera tracks down to ENVELOPE. It is addressed to De La Gardie.

C.S. Wraxall licks his lips and concentrates on the writing. He looks a haggard and broken man.

C.U. Page of letter (text).

> My Dear Count,
> I have to travel to Skara for a few days. I shall be staying at the Three Crowns.

C.U. Wraxall's face. It wears a look of despair.

Dissolve to:

EXT. RABACK STATION AND PLATFORM.

M.S. Steam obscures lens. Camera pulls back a little to show piston rods of engine.

L.S. Angled along platform of small country station. PORTER, one or two PASSENGERS. Name board says: RABACK. Wraxall comes rapidly along platform toward camera, looking sharply about him.

M.S. A MAN IN A CLOAK goes round a corner of the station building.

M.S. Tracking shot of Wraxall's face. He looks agitated.

L.S. Platform, Reverse angle. Man in cloak comes toward camera. He is sturdy country farmer. He opens CARRIAGE door and gets in. Wraxall comes up toward camera. He looks relieved and gets into same carriage.

INT. RAILWAY CARRIAGE.

M.S. Wraxall gets in and sits down in seat farthest from camera at WINDOW. Countryman is already seated in foreground. He says something in Swedish to Wraxall, who smiles and mumbles something inaudible in return.

EXT. RABACK STATION AND PLATFORM.

L.S. The train at platform. The porter waves it off. It starts to move, steam billowing across platform. The train pulls out, away from the camera. There are only about three carriages.

C.S. Driving wheels of engine pass camera.

M.S. Tracking shot of Raback station name board.

INT. RAILWAY CARRIAGE.

C.U. Wraxall's relieved face. He stares out of window as TRAIN WHISTLE sounds mournfully.

EXT. COUNTRYSIDE.

L.S. The train rushes through the countryside.

INT. RAILWAY CARRIAGE.

M.S. Camera angled down on to swaying feet of countryman. Slow pan up to Wraxall, silent and absorbed in his corner.

EXT. COUNTRYSIDE.

M.S. Countryside sliding past window.

INT. RAILWAY CARRIAGE.

C.U. Wraxall. He is lost in thought. He turns his head to look out of window.

EXT. LONELY CROSSROADS.

L.S. Moving shot of countryside. SMOKE obscures frame, then clears to reveal a lonely crossroads. The roads intersect in a bleak manner and run straight across the plain. There are LOW HEDGES. Brief, oblique shot.

INT. RAILWAY CARRIAGE.

C.U. Wraxall's expression is now intent. He moves toward window.

EXT. LONELY CROSSROADS.

L.S. Camera from train window slowly moves across crossroad. There is something which looks like a GIBBET but which is probably a sign-post. Two figures are standing motionless at the crossroad; one is tall, with a cloak and wearing a broad-brimmed hat. He leans on a staff. The other is short, in black, and wearing a hood.

INT. RAILWAY CARRIAGE.

M.S. Wraxall sinks back on to his seat in desperation. He produces a HANDKERCHIEF and mops his face.

M.S. Reverse angle. Countryman looks at him curiously.

EXT. TRAIN WHEELS.

C.U. Train wheels roar as they pound at the rails.

C.U. Steam whirls past lens as whistle blows.

EXT. COUNTRYSIDE.

L.S. Crossroads and hedges; no figures are now visible.

The train rushes through countryside.

INT. RAILWAY CARRIAGE.

M.S. The carriage. Wraxall is slumped in the corner with his hands covering his face.

EXT. SKARA STATION AND PLATFORM.

L.S. Train pulls into slightly larger station.

INT. RAILWAY CARRIAGE.

M.S. Carriage. Stranger starts getting his belongings together in foreground. Station name board slides past window: SKARA.

M.S. Wraxall pulls himself together. He takes his VALISE off the RACK, glad of the bustle.

EXT. SKARA STATION AND PLATFORM.

L.S. Angled toward end of train. PEOPLE are coming down platform to board it, passing in front of camera.

M.S. Closer to Wraxall's carriage. The countryman gets down with his PACKAGES, looks back curiously into interior and then passes camera. Slight pause and then Wraxall gets down on to platform with his valise. He looks uncertainly about him.

M.S. Station buildings, with bustling passengers passing to and fro. STEAM drifts across frame.

L.S. Wraxall walks away from camera, down the platform toward the exit, a lonely and dejected figure.

Fade.

NARRATOR (V.O.)

Poor Mr. Wraxall! He now realised for the first time that he had set in motion forces he could not control. His one instinct was flight.

INT. STUDY, RABACK MANOR HOUSE.

L.S. Interior of study at Raback Manor House. De La Gardie is reading Wraxall's letter with slowly mounting alarm. Pastor Moeller is seated near the Count's desk with a GLASS OF WINE at his elbow. He waits with interest for the Count to finish reading. De La Gardie looks up with a CRACKLING of notepaper as he turns to Moeller. It is late afternoon, with a dying sun coming through the windows.

MOELLER

Not bad news, I hope?

DE LA GARDIE

It is Mr. Wraxall. He is in some kind of trouble.

(he gets up from his desk)

He is staying at the Three Crowns in Skara. I must go to him.

C.S. Pastor Moeller. His eyes flash and he is about to say something, but bites back his words. He turns his head to look at the LARGE CLOCK on the MANTEL SHELF behind him.

MOELLER

I will come with you. But it is six o'clock already. We had best dine now. There is no further train until eight o'clock this evening.

C.S. De La Gardie. He looks more reassured. He moves toward Moeller.

DE LA GARDIE

Thank you, Pastor.

L.S. From behind desk as De La Gardie walks over to Pastor Moeller. The two men walk out the far door.

Dissolve to:

EXT. THE PARSONAGE, SKARA.

L.S. A large white church. It is dusk. Wraxall comes along the path. He looks distraught and agitated. He looks behind him once or twice, nervous at the waving branches of the trees in the RISING WIND. He turns away from the church and goes up toward the PARSONAGE, the camera panning slightly with him.

L.S. The parsonage, a low, severe-looking building, all black and white. Wraxall comes into frame and hurries toward the building.

M.S. The parsonage DOOR. Wraxall almost rushes at it and hammers with the heavy iron KNOCKER. Shadows of tree branches dance wildly on the door. He looks furtively about him.

C.S. The door edge. Shadow grows as the door opens. The heavy, white features of a middle-aged WOMAN compose themselves from the dusk. She looks uncomprehendingly at Wraxall.

M.S. Taking in Wraxall, front of house and Housekeeper.

WRAXALL

(brokenly)

I must see the Pastor. It is very urgent.

C.U. The housekeeper stares uncomprehendingly at Wraxall.

C.U. Wraxall. He licks his lips and tries again.

WRAXALL (CONT'D)

Do you understand? I must see the Pastor!

(he looks about him in the dusk wildly)

C.U. The woman smiles, though she obviously understands nothing. She starts talking rapidly in Swedish.

M.S. Wraxall listens uncomprehendingly. The long path is seen over his shoulders. Something like two figures in black pass across the end of the path, as though going toward the church. Wraxall does not notice them.

C.U. The woman goes on speaking.

C.U. Wraxall. He makes a pathetic effort to understand what the woman is saying.

WRAXALL (CONT'D)

The Pastor... I have to speak to him...

M.S. From behind Wraxall, shooting toward the housekeeper and the parsonage door. She speaks more harshly, evidently considering Wraxall stupid. She makes large gestures with her hands, pointing down the path. Wraxall understands at last. His shoulders slump.

C.U. Wraxall's wretched face.

WRAXALL (CONT'D)

Oh, away, eh? Thank you very much.

M.S. Wraxall raises his hat.

WRAXALL (CONT'D)

I am sorry to have troubled you.

The housekeeper smiles. She gives him a little bow and points again, regretfully, down the path up which he has come. Wraxall comes toward the camera, dejectedly.

C.S. The housekeeper looks regretfully after him. She closes the door slowly.

L.S. Path taking in parsonage in background and part of church adjoining. Wraxall, a crushed, broken figure comes toward camera. He hesitates and then looks toward church. The camera pans to left to take in church as he quickens his footsteps and goes toward it.

M.S. Church from nearer angle. Wraxall enters frame and almost runs toward the porch. He opens the porch door and enters. Hold for moment.

INT. SKARA CHURCH.

L.S. Darkling interior of church. The door opens to admit a pale oblong of light. Wraxall comes hesitantly toward the camera.

L.S. Reverse angle. WHITE WALLS, NARROW WINDOWS, a profusion of CARVING about the ALTAR. A few CANDLES are burning, throwing long shadows. Wraxall advances hesitantly toward the altar.

M.S. Wraxall comes toward camera. He looks tired and wretched, a beaten man.

M.S. Side on to altar. Wraxall comes into right-hand frame, looks hesitantly about him.

C.U. Wraxall's face. He looks up.

C.U. The CARVED FACE of Christ on the cross.

C.U. A saint in a CARVED FRIEZE.

C.U. A bearded saint in an OIL PAINTING which hangs on the wall.

L.S. From above the altar, looking down on Wraxall.

M.S. Wraxall slowly sinks to his knees in a praying position. He looks fearfully at the shadows.

C.S. The candle flames flicker in the draught.

C.U. The face of Christ appears to smile in compassion.

C.U. Wraxall's face. He is silently crying.

L.S. High up, from rear of church, looking down at the tiny kneeling figure in front of the altar. Hold and then

Slow fade.

EXT. THREE CROWNS INN, SKARA.

M.S. Hotel exterior. Night. An owl HOOTS somewhere and a CARRIAGE drawn by a HORSE passes in the distance.

C.U. Inn SIGN BOARD, moving in the NIGHT WIND, with branches throwing shadows across. THE THREE CROWNS.

Dissolve to:

INT. THREE CROWNS INN, SKARA—WRAXALL'S ROOM.

M.S. A dark-panelled room in inn, severely, yet comfortably furnished. Wraxall is sitting at a TABLE writing.

C.U. Wraxall's absorbed face.

INT. RAILWAY CARRIAGE.

M.S. Night. Moeller and De La Gardie are sitting side by side. Carriage is full of PEOPLE. Both men look anxious.

EXT. STREETS OF SKARA TOWN.

M.S. Narrow streets outside the inn. The camera slowly pans across ROOFTOPS. Only a few faint FOOTFALLS sound.

INT. THREE CROWNS INN, SKARA—WRAXALL'S ROOM.

M.S. Wraxall pauses in his writing.

EXT. STREETS OF SKARA TOWN.

L.S. The narrow streets. A tall figure in a cloak and a short one with a hood are faintly glimpsed, walking toward camera. There is no sound of footsteps.

EXT. SKARA STATION AND PLATFORM.

L.S. The platform at Skara station. The TRAIN has stopped and STEAM blows across the platform. De La Gardie and Moeller hurry toward the camera with worried expressions.

EXT. THREE CROWNS INN, SKARA.

C.S. Three Crowns signboad. Camera pans slowly up to light of Wraxall's room.

INT. THREE CROWNS INN, SKARA—WRAXALL'S ROOM.

M.S. Wraxall is sitting listening in a tensed position. He slowly relaxes and starts writing again.

EXT. ROOFTOPS SURROUNDING THREE CROWNS.

L.S. Roof-tops. A long shadow slowly falls across the tiling.

EXT. STREETS OF SKARA TOWN.

L.S. Street. De La Gardie and Moeller hurry along, looking about them.

EXT. ROOFTOPS SURROUNDING THREE CROWNS.

L.S. Roof-tops. Night SOUNDS OF INSECTS. The camera moves forward and then freezes. Short, squat shadow moves across tiles.

EXT. THREE CROWNS INN, SKARA—ENTRANCE HALL.

M.S. Entrance of Three Crowns hostelry. Camera pans down to ground level as De La Gardie and Moeller hurry toward the main doors. Moeller goes on ahead.

EXT. ROOFTOPS SURROUNDING THREE CROWNS.

M.S. Camera tracks steadily toward the lighted WINDOW of Wraxall's room.

INT. THREE CROWNS IIN, SKARA—ENTRANCE HALL.

M.S. Moeller and De La Gardie are talking with LANDLORD. Landlord gestures toward the STAIRCASE.

INT. THREE CROWNS INN, SKARA—WRAXALL'S ROOM.

M.S. Wraxall pauses in his writing. A low, chilling LAUGH sounds from outside the window.

C.S. Wraxall turns his head sharply toward the window.

M.S. The window, night beyond. Shadows fall across the sill.

M.S. Wraxall jumps to his feet.

INT. THREE CROWNS INN, SKARA—STAIRCASE.

M.S. De La Gardie and Moeller, with landlord ahead, are walking up the stairs.

INT. THREE CROWNS INN, SKARA—WRAXALL'S ROOM.

M.S. Camera framing window of Wraxall's room with Wraxall standing stiff with horror beyond. Camera tracks swiftly in through window to Wraxall's face.

INT. THREE CROWNS INN, SKARA—CORRIDOR.

M.S. De La Gardie, Moeller and landlord freeze as SCREAM AFTER SCREAM echoes throughout the hotel.

INT. THREE CROWNS INN, SKARA—WRAXALL'S ROOM.

B.C.U. Wraxall's screaming face as the shadows fall over it.

INT. THREE CROWNS INN, SKAKA—CORRIDOR.

M.S. De La Gardie and others rush to the door of the room. Landlord finds it locked, pounds on it. The screaming stops.

DE LA GARDIE

Are you there, Mr. Wraxall? Why don't you answer?

M.S. Different angle. There is silence from the room now. The landlord listens at the panel, turns serious face to the others.

L.S. Corridor is filling with PEOPLE. Moeller has a PRAYER BOOK in his hand, is intoning from it silently. De La Gardie looks white and strained. Landlord takes MASTER KEY from WOMAN who hurries up. Camera pans to door as he inserts key in lock.

INT. THREE CROWNS INN, SKARA—WRAXALL'S ROOM.

L.S. Interior of Wraxall's room, facing door. It begins to open inward.

M.S. Landlord's frightened face, De La Gardie, Moeller and others glimpsed beyond.

L.S. Window. Curtains rustle in the wind; a shadow seems to slip away across the sill and out over the rooftops. Camera slowly pans right from window and down to floor. Wraxall's shoes are seen lying on boarding.

M.S. Landlord and others, looking downwards across the room.

C.S. Continue pan to end on Wraxall's spectacles on floorboards.

M.S. Landlord shifts his gaze.

C.U. Moeller, trying to look over his shoulder.

C.U. De La Gardie trying to see.

C.S. Camera pans over floor. There are splashes of blood scattered about.

C.U. Landlord's appalled face.

C.U. Wraxall's SEVERED FINGER lying on the floor.

M.S. Swift track in to De La Gardie's stunned face as someone SCREAMS.

M.S. Whole group. Landlord turns away retching as group breaks up.

Dissolve to:

EXT. STREETS OF SKARA TOWN.

L.S. Shadows pass across the moonlit streets of the little town.

L.S. A small street. PASSERBY stops, startled as horrible LAUGH echoes along the winding way.

M.S. Passerby looks about him.

L.S. Oblique shot of side-street. Two shadows crawl along it, elongated.

Dissolve to:

EXT. LONELY ROAD.

L.S. Two figures, one tall, one short, are walking with strange, jerky gait, in the middle distance along a lonely road outside the town.

Dissolve to:

Extreme L.S. Tall figure in cloak and with broad-brimmed hat stands silhouetted on a knoll in moonlight. Hold, then slowly

Fade.

NARRATOR (V.O.)

People in Skara still remember how a strange-looking gentleman came to the Three Crowns some years back. And how that one evening he was found dead in his room. There was an inquest and the jury that viewed the body fainted to a man. The verdict was a visitation of God. No glimmer of light has ever been thrown, or could be thrown, on the mystery of Mr. Wraxall's death.

I. UNPUBLISHED WORKS

Includes an alphabetical listing of unpublished, incomplete and unrealised books, stories and other works by Basil Copper

(*"Sometimes books or novellas are written to order for publishers or editors, who sometimes go bankrupt or lose their jobs respectively, and then such material sometimes lies in limbo, on occasion for decades. But they have a habit or reappearing after the most astonishing intervals.*

"One novella of mine was finally published twenty-five years after it was written; a major novel of mine appears in America this autumn after a time-lapse of ten years.

"I recently rediscovered a macabre tale of mine I penned at the age of sixteen! It is to be published this year bearing the copyright line of 1940! So my advice to any aspiring writer is—never throw anything away! You may be sitting on gold. Short stories, in particular, if they are good enough, are anthologised in hardback and paperback time and time again, for excellent fees these days." BC.)

I1 **THE CURSE OF THE FLEERS**

a. __________, unpublished.

[Note: Although a bowdlerised version of this macabre novel appeared on both sides of the Atlantic in 1977, Basil Copper does not consider that this book has been properly published yet.]

I2 **DEATH TO THE INVADERS**

a. __________, unpublished.

[Note: Juvenile World War II novel about German paratroopers invading Britain. This apparently no longer exists.]

I3 **THE GRIMOIRE OF ASTAROTH**

a. __________, unpublished.

[Note: Macabre short story that exists as an incomplete first draft manuscript (8pp) dated May 19, 2005.]

I4 THE HORROR ON PLANET X

a. __________, unpublished.

[Note: Science fiction novel in manuscript form (209pp), written March-April, 1974.]

I5 KEEP WATCHING THE SKIES

a. __________, unpublished.

[Note: Science fiction short story, apparently never completed.]

I6 LETTERS TO ARKHAM: CONVERSATIONS WITH AUGUST DERLETH AND JAMES TURNER

a. __________, unpublished.

[Note: A proposed non-fiction book collecting Basil Copper's correspondence with August Derleth and James Turner at Arkham House, Publishers.]

I7 MARK OF THE BEAST

a. __________, unpublished.

[Note: Proposed anthology based around different creatures and featuring stories by William Hope Hodgson, H.P. Lovecraft, Sir Arthur Conan Doyle, "Saki" (H.H. Munro), A.E. Coppard, Jack London, Guy de Maupassant, H.A. Manhood, Bram Stoker and Ambrose Bierce, along with Basil Copper's own tale 'The Spider' (G82). A three-page 'Foreword' was written April 25, 1994.]

(*"The genesis of this book began following my reading a collection of short stories by that wonderful pre-1914 writer of ghost stories with sea backgrounds, William Hope Hodgson. The particular collection,* Deep Waters, *had been edited and published by my old friend the late August Derleth under his celebrated Arkham House imprint.*

"Two stories in the book particularly impressed me: 'From the Tideless Sea', which told of a horrifying plight of a man, his wife and child imprisoned for years in a ship held fast in the Sargasso Sea; and 'The Thing in the Weeds'. These were tales of terror par excellence *and yet, far from having an extra-terrestrial basis they were found at the end to have a natural explanation.*

"Curiously, this did not diminish them as tales of horror and I wondered whether a collection of stories with the terrors based on animal, insect or reptile might not have an equal fascination for a wider audience." BC.)

I8 THE RESTLESS DEAD

a. __________, unpublished.

[Note: Following 'The Great Vore' (G51), this was intended to be the second of three macabre novellas featuring psychic investigator Professor Alden Kane. Two pages of notes exist, written May 19, 1975 and January 6, 1976, with the dedication "For James Turner, Who Made Me Write More of Professor Kane's Adventures".]

(*"Plot points: Blundalstone on bleak Norfolk coast: Professor Kane there just before Christmas. Christmas to be connected with the plot: man who passes him is dead pastor, Isaac Roche? Leaves no footprints in the snow. Read antiquarian article, which gives him a clue: man (local curator) comes to him with terrifying problem. Kane has awful dream which is evidently connected. Telegraphs for his old friend Edwards. They brave terror to solve the problem and in end stay over to celebrate New Year.*

Scene in museum with old book. Written by Holubar Quint . . .

Colonel Manson comes in again: helps. Several people die:

The Brethren (the living in Blundalstone aid the dead in their rituals) did not like Toliver's recent series of articles: he fears for his life with the publication of the latest, the previous day. Attempt on his life prompted his call on the professor.

Quint a 15th century wizard: said to have been burned on a mound outside Blundalstone." BC.)

I9 **THE SOLAR PONS COMPANION**

Contents: Plots of the Stories, Characters in the Novellas, The Sayings of Solar Pons.

a. __________, unpublished.

[Note: Non-fiction lexicon in manuscript form (100pp), written 2000-05.]

I10 **UNTITLED CINEMA BOOK**

a. __________, unpublished.

[Note: Since the 1960s, Basil Copper has written hundreds of pages for a proposed volume on Cinema History and Film Collecting that has never been completed.]

I11 **UNTITLED LOVE AFFAIR NOVEL**

a. __________, unpublished.

[Note: A proposed semi-autobiographical novel about a middle-aged man in the late 1950s who finds a bundle of old love letters and sets out on a pilgrimage to Yorkshire. Intended to be around 300 pages long, with each alternating chapter told in flashback, a first draft of twenty pages exists (page 5 is missing) consisting of a Prologue and Chapter One.]

DARKNESS, MIST & SHADOW
THE LANDSCAPE OF FEAR

[Note: 'Guest of Honour's Lecture' at Fantasycon 3, The Imperial Hotel, Temple Street, Birmingham on February 26, 1977.]

AS H.P. LOVECRAFT so aptly remarked long ago, "The oldest and strongest emotion of mankind is fear, and the oldest and strongest kind of fear is fear of the unknown".

This seems as good a definition as any to me and admirably sums up my thesis here today at your British Fantasy Convention. I have called my speech, "Darkness, Mist and Shadow–The Landscape of Fear" and I shall have occasion to stress it more than once as I develop my theme.

I find it necessary to state at the beginning that fear is one of the chief weapons in my armoury as a macabre writer; I call myself a macabre writer and it is perhaps as well from the outset to delineate my terms of reference. I work, as well as I am able, within the tradition of Lovecraft in his macabre mood; of M.R. James in particular; of Poe, of course; and all those other writers, great and upcoming, who then devoted themselves to the small but distinguished niche of the macabre.

The macabre, which means, for me, the horror story, the *conte cruel*, the ghost story proper; the tale of vampirism or lycanthropy; the narrative which sustains the mood of pure terror. There is a rich field to be mined here but I must make it clear that fantasy, as your members understand it, has a large part

by Basil Copper

of it. Fantasy and poetry, for the two are intertwined.

Not the fantasy of heroic cities, dragons, strange adventurers and lands beyond the stars; but the fantasy and poetry of bizarre occurrences which are, in my own particular case, rooted fairly closely to the familiar bounds of this earth.

That is not to say, of course, that I have not the highest respect and admiration for those fine and skilled writers who work within the realm of pure fantasy, among which Mr. Michael Moorcock must stand supreme on this side of the Atlantic, just as Ray Bradbury does in the United States.

And it would be untrue to say, again in my own case, that I have not indulged in such dreams; as in my novella 'The Flabby Men'; in such tales as 'The Janissaries of Emilion'; or in my fantastic novel *The Great White Space* which, however, owes more perhaps to Conan Doyle than the kings of sword and sorcery. I find it necessary to state this emphatically in order that I do not stand before you under false pretences.

So when I speak of the macabre would you please read fantasy; and when you talk of fantasy I will make a similar substitution!

Whether one is dealing with fantasy in the swords and sorcery tradition; or in the macabre as defined in my opening remarks, one is faced with two problems. The purely technical; the nuts and bolts from the writer's point of view of getting one's ideas across to the reader with the maximum impact. And the reader's in becoming absorbed in the story; of the suspension of disbelief, which is tied up with the writer's skill, and also allied to the oldest puzzle of all. Why does the writer in the first place want to write something so strange and horrible? And why does the reader wish to read something that at its best will terrify and appal?

The oldest ghost story in the world is supposed to run—'In the dark a man reached out for the light. A light was put into his hand'—That still carries a genuine *frisson* today for those with the imagination to be thrilled by the skill of a genuinely good macabre writer. There are few of them about; there never have been many. And those of us active in the field today owe a great debt to the giants of the past.

I will first attempt to answer my questions already posed. The good macabre/fantasy writer will also want to write out of the best in himself because he enjoys reading the work of others in the same field. The short story or novella, particularly in the macabre/fantasy genre, is one of the most difficult things in the world to do well. No writer worth his salt is ever satisfied with what he has achieved; I know I never am. And the writer of the macabre has first to convince the sceptical reader that what he is reading actually happened; and then, on top of that, evoke a thrill or a shudder in a man who is sitting in a comfortable armchair in a clean, well-lighted place, perhaps with a television set competing for attention in the corner. An almost impossible task, but the best in the field have done it—occasionally.

For the writer in this genre there is nothing like really getting home to the reader. It would be impertinent and immodest of me to state here that I have sometimes done that but, occasionally, to judge by the (few) letters which reach me, it would appear so; it is extremely gratifying for the man who earns his living by the written word.

Speaking now as a reader of other men's work, it is equally difficult to answer why one should deliberately sit down to read something which may frighten, awe or even terrify! It is almost as strange as being asked why one writes such things. I think pleasure comes high on the list, of course; the same pleasure one would get from reading something well-written in any other branch of literature. I leave out the critics here; the most Olympian would deny that the macabre/fantasy is literature at all, but I think we can ignore them. Those perceptive critics who do give generous space to the genre, particularly the large-hearted gentlemen in America, take this branch of literature very seriously indeed and more power to them.

In addition to pleasure I feel must come a perverse liking for living dangerously, even at third-hand. Perhaps the luxurious sensation of reading about someone in mortal peril while the reader himself sits comfortably in his chair, a glass of whisky at his elbow. And, of course, able to escape from the menacing horror in an instant by withdrawing his eye from the page.

As well as these other motives must come the purely sensuous pleasures that sensitive people get from poetry and imagination; they certainly get this in great measure from the works of pure fantasy and they should get it in the well-wrought tale of terror.

For the writer then, the achievement of a difficult task and major obstacles surmounted; for the reader, a vicarious thrill and the abiding pleasure of a story which is plausible, well-written and preferably of an agreeable flavour, even if only of the dank chill of an old house; or a mist-wreathed churchyard. Darkness, Mist and Shadow—The Landscape of Fear.

I have not yet touched upon one aspect very important to me personally; restraint. I feel—and I may be challenged on this—that restraint is of the essence. Those writers and creators at the cheaper end of the field—like the makers of so many of the Hammer horror films—seem to me to be so wrong in wallowing in severed heads and loose eye-balls. Their production values seem to me to smack more of the abattoir than of artistic endeavour; and more often than not raise laughter and disgust rather than the chill of genuine fear.

A shadow on the wall; fingers at the window; a low laugh in the gloom of the staircase. These things, to my mind, are more truly terrifying than the blood-stained corpse and the hundred-piece orchestra. Then the reader—or the spectator's—imagination will do the rest and the result will be far more satisfying, both to creator and the partaker of the finished product. For it appears to me—reinforced by the intelligence on the faces before me—that this business of writer and reader is a very live and real relationship. One could not exist without the other; and if the writer is doing his job properly, and the reader bringing a degree of concentration and intelligence to his task, the result is as satisfying and pleasurable as a good game of tennis. Without these criteria, it is exactly like playing without a partner.

I have spoken of restraint; of darkness, mist and shadow, in the genre of the macabre/fantasy. Let me give some specific examples. The greatest master of the pure ghost story, M.R. James, with his forbidding, inexplicable events set on the bleak Norfolk coastlines of the turn of the century is an admirable model and can spell out a good deal of outstanding examples for today's practitioner and reader.

It was James who conjured up one of the most horrifying demons ever evoked on paper with the man, in his famous story 'Casting the Runes', who reached under his pillow for his watch. James says, "What he touched was, according to his account, a mouth, with teeth, and with hair about it, and, he declares, not the mouth of a human being".

Just think about it; there's a deadly restraint in the sentence. The film which emerged from this story some fifty years later, *Night of the Demon*, though excellent—certainly one of the best things of its kind—did not attempt to re-create this sequence. Wisely, perhaps, one might think.

It was Dr. James again in perhaps his most celebrated story, 'Oh, Whistle and I'll Come to You, My Lad', who evoked possibly the most horrifying—and certainly one of the most novel—familiars in all macabre literature when an "intensely horrible face of crumpled linen" was thrust into his own. The bedclothes had risen of their own volition, you will remember, and formed an entity which attempted to push the hero, a mild, ineffectual scholar, out of his hotel window. It was James too who invented the face with cobwebs across the eye-sockets. He is certainly one of the great masters, and the writer who follows his example—without copying him, of course—will come up with some wonderful ideas. The mundane made horrible.

The workings of Dr. James' remarkable mind are apparent in a piece he wrote for a collection of his stories published as far back as 1904. In it, he said, "There may be possibilities, too, in the Christmas cracker, if the right people pull it, and if the motto which they find inside has the right message on it. They will probably leave the party early, pleading indisposition; but very likely a *previous engagement of long standing* would be the more truthful excuse.

"In parenthesis, many common objects may be made the vehicles of retribution, and where retribution is not called for, of malice. Be careful how you handle the packet you pick up in the carriage-drive, particular if it contains nail-parings and hair. Do not, in any case, bring it into the house. It may not be alone . . ."

These are ideas which still hang vibrant in the air more than seventy years since he first put them to paper. So vivid and imaginative; so horrifying if handled by the right pen. The restraint of the idea combined with the imagination of the reader . . .

I think this strange, sly and curious mind was what gave me such pleasure when I scripted James' celebrated 'Count Magnus' for a television film which is still pending, more than two years after the acceptance of the script.

Like most macabre/fantasy writers, Dr. James, who was a former Provost of Eton College. had a lively sense of humour. In the same essay he remarked, "Dots are believed by many writers of our day to be a good substitute for effective writing. They are certainly an easy one. Let us have a few more . . .

"Late on Monday night a toad came into my study . . ."

Darkness, Mist and Shadow. Restraint and artistry in the writing which composes the landscape of fear. People often ask how such ideas as those of Dr. James I have just outlined come about. The creative process is a strange one and often stems from mundane things. Certainly Dr. James' witch-haunted mind apparently felt little strain in effortlessly capturing such brilliant demons out of the darkling air.

In my own case, though I have no shortage of story ideas—my commonplace book on the macabre contains something like two hundred—it is simply the time needed to get them all down, especially when confronted with the press of other tasks. Life is simply not long enough; in my own case I write in at least ten different genres, including two contrasting styles in the detective field; private-eye Los Angeles; and the more leisurely world of yesterday in the atmosphere of the new Solar Pons series which I am currently creating for Arkham House.

But I think it would be not uninteresting to mention some of the macabre ideas which have come to me from time to time and their genesis as near as I can recollect. Dennis Wheatley, as a young writer before the war, in a Foreword to one of a famous series of books, A *Century of Horror*, put the definition of horror in a striking manner.

In a perceptive article he wrote, "Imagine me, then, working away on a summer night at my table near the open window. I have just finished a chapter of a new book perhaps. I sit back with a sigh of satisfaction and then—to my unspeakable horror—I see a cobra seated on the sill of the open window.

"It is coiled like a huge watch-spring—its broad flat head reared up and swaying slightly, its forked tongue flickering in and out with incredible rapidity as it fixes me with its beady black malignant eyes.

"What I should do in such circumstances I cannot attempt to say. I should certainly be scared out of my wits but, after the first shock, reason would probably return to me. I should at least be able to explain its presence by the theory that it had escaped from the Zoo, which lies only half-a-mile away and, if the brute attacked me, attempt to defend myself by frantic slashings with the hooked pole by which I open and shut my windows.

"Now let us suppose that, instead of a snake I saw a slim white hand, which had no arm or body behind it, reaching in at the window to grab me.

"My 'horror' would be fifty times as great, for no material theory could explain that! I should be too paralysed with fright to do anything at all and have no faith whatever in the idea that a well-aimed blow with my window-pole could destroy such a menace".

This Introduction, which I had read before the last war, flickered across my mind one night about three years ago. I live in a house built round a courtyard. Though it is in a town, it is in a small valley entirely secluded and might be miles in the countryside. One night, in the depths of winter, I was preparing for bed. It was about 1:30 a.m. and I had just locked the front door. I was standing motionless, having shot the bolt, thinking about nothing in particular, when there came a terrible scream from the courtyard outside.

It was a considerable shock and I remained rooted to the spot. Reason told me that the scream was that of a fox; there is a small colony nearby and a vixen screaming is a familiar sound during the winter months. Sure enough, though I did not see the animal, there were the paw-prints on the frosty ground the next morning. But the shock of that moment had started a train of thought which gave me a bizarre idea for a story.

Supposing, for example, there had sounded at that dead hour of the night, the high wail of a young child. And at almost the same instant the pink arm of a baby had been thrust through the letter-box. The shock of such a moment in real life might be enough to turn the mind and would certainly be unexplainable. The idea is filed; whether I shall ever use it is another matter.

One might ask the writer of the macabre and fantasy field if he himself believes in the supernatural. I keep an open mind and it certainly helps. As a newspaper editor many years ago I was concerned with events in a reputedly haunted Kentish inn. The occurrences during the long nights we waited and watched literally made my hair stand on end more than once. A single example will suffice. What could explain the phenomenon when the pictures on the walls of the hotel dining room began to dance about and stand out from the walls, moved by some invisible force? Investigators of the paranormal talk and write glibly about poltergeists and attempt to rationalise away such happenings. I shall never forget my own horror when attempting to hold one of the pictures flat against the wall and feeling the insidious, unseen force inexorably putting pressure against my finger-tips. All the terror evoked in Guy de Maupassant's 'The Horla' is in such a moment, and it actually happened.

So I suppose such experiences tend to colour one's own writing; they should certainly add credibility to the fictional horrors. I hope they do so in my case. Though I have not consciously subsumed this event into one of my stories, others have been. An incident when I was staying at a friend's villa in Egypt during the war gave me a lasting horror of spiders. This Alexandrian spider was a tremendous brute and my fear at finding it near my bed when preparing to retire for the night was only paralleled by the delight of my host, an eminent Egyptian parasitologist who promptly chloroformed it for his collection! Traces of that war-time experience got into one of my nastiest stories, 'The Spider'.

I am, as some of you probably know, a keen collector of historic film material and have one of the largest collections in private hands in this country. Among my silent films are three MGM Lon Chaney melodramas and Wiene's famous *Cabinet of Dr. Caligari*. I often watched this weird and sombre drama on my own late at night and the huge close-ups of Conrad Veidt as Cesare eventually found expression on paper in my story 'Amber Print', which Christopher Lee was kind enough to say is one of his favourite macabre tales.

There are many similar and perhaps less obvious ways in which a writer's imagination is set to work. Another came from a friend's detestation of television. This is pencilled in my commonplace book as 'The Man Who Hated Television'. This envisages a man whose wife purchases a set against his protests. His private life is disrupted and he retreats into his shell, while his partner sits in front of the set year after year, entranced and oblivious of his needs. One night his wife disappears. After a while neighbours burst in; the police eventually find the large TV set in the lounge has been disembowelled. The works have been removed; in their place are the dismembered and decomposed remains of his wife. The story would end with a scene in an asylum. All the inmates are sitting in a large room watching television. All except one. He sits motionless in a chair slewed round, his back to the screen.

Even this latter detail stems from reality; another traumatic experience when I once ill-advisedly gave a programme of historic films at an old people's home. One ancient lady with a green eye-shade sat grimly with her back to the screen and refused to budge, asking her neighbour to give her a running commentary on the events taking place on the screen behind her. The horror of that moment remains with me yet. But something might be made of the story idea which I have just outlined.

My own preferences in macabre/fantasy, as I have amply made clear, are not for shining cities or gleaming domes, however seductive they may be, but for events which took place just the day before yesterday. Horrific, bizarre and strange events which might have happened in that shifting borderland which lies just beyond the twilight zone. Darkness, Mist and Shadow again. The Landscape of Fear.

I seem to be getting rather deeply into quotation by way of illustration, but I cannot resist another which to my mind is one of the finest definitions of the criteria necessary to the macabre and fantasy writer.

It comes from Carl Dreyer, one of the world's greatest film directors and the author of one of the ten best macabre films of all time, *Vampyr*. No doubt the remarks may be familiar to some of you, but I feel they are worth repeating in the context of my speech.

Dreyer said, when he was making this strongest of all films in 1932, and on an occasion when he had gathered all his cast and technicians together, "Imagine that we are sitting in an ordinary room. Suddenly we are told that there is a corpse behind the door. In an instant the room we are sitting in is completely altered; everything in it has taken on another look; the light, the atmosphere, have changed, though they are physically the same. This is because *we* have changed and the objects are as we conceive them. That is the effect I want to get in my film".

The Corpse in the next Room. What a wonderful simile. That is the effect which every good and painstaking writer in this genre is out to achieve. At their best what Poe and Lovecraft and Arthur Machen and Algernon

Blackwood and L.P. Hartley and M.R. James and E.F. Benson and a few others in the past have done and what we, of a new generation, are trying to effect in our turn. Their way, I appreciate, is not everyone's way, but it is mine and I am firmly committed to it. Naturally, I can speak only for myself. As I have said, my own preferences are not for shining cities or mystical domes, though I can understand and appreciate the fine qualities in this type of story and indeed, briefly evoked my own fabled city in 'The Janissaries of Emilion', a place of which the narrator dreamed.

But those who have chiefly influenced me in my subjects, not my style, are Poe, Lovecraft and M.R. James in their most macabre moods. Poe at his most gifted as in 'The Fall of the House of Usher'; Lovecraft, in those tremendous stories 'The Rats in the Walls'—surely one of the greatest tales of horror of the twentieth century—and my own personal favourite, 'The Whisperer in Darkness'; and M.R. James in the great body of his work which, like the best horror films, merely sketches in the intangible horrors of Darkness, Mist and Shadows.

Subtlety cannot be over-estimated in the genre. When inventing blasphemous books, a ploy much favoured by macabre writers, I would advise: don't invent too many. Be circumspect. My only addition to the range is *The Ethics of Ygor* in my *The Great White Space*. Aficionados will recognise the genesis: the malignant companion of the Monster as personified by Bela Lugosi in *Son of Frankenstein.*

Similarly, when coining fictitious place-names, my own preference is for something that sounds like rotting fungi being crushed underfoot. Lovecraft caught it perfectly with "Pnath", if that be the correct pronunciation. I think I did the trick with my own coinage, In the Vale of Croth in 'The Grey House', but it is, after all, something the reader will judge for himself.

It is essential, when writing a macabre story, to have it well plotted and structured, and to clothe it with atmosphere as one proceeds with the actual writing. My own working method is to have the denouement or ever the final sentence clearly in my mind when beginning work.

Like most professional writers my working life is sane and orderly. I rise at eight o'clock each day and am at my desk by 9:00 a.m. I begin with business letters and personal correspondence, proof-reading and so on, and am usually at my creative work by midday. I break for lunch at one o'clock and afterwards walk up to the town to post my letters and visit shops. This half-an-hour is one of the principle things which retains my link with sanity. By three o'clock I am back at my desk and work steadily on until half-past five or six o'clock like any businessman. This is my invariable routine and I never work beyond six o'clock.

I work as a touch-typist, typing my books directly on to the paper, with a minimum of alteration or revision, and a secretary follows me up, chapter by chapter. I usually work on four or five books at once, giving an hour to each, as I find I become stale otherwise, and sometimes write as many as eight books a year, most of them in the non-macabre field, of course. The difficulties in the latter genre can hardly be over-emphasised when I mention that a macabre novella of mine being published this spring was written twenty-five years ago!

Unfortunately, one has to diversify and have a large output if one is to earn enough money and survive. I could mention some horrific stories of publishers who do not pay and agents who operate just this side of fraud on both sides of the Atlantic, but that is a macabre story in real life and I will not bore you with it here.

It is sad to reflect that most of the great writers in the macabre/fantasy field of the past hundred years have had tragic lives and have often died in poverty. One thinks of the classic examples of Poe and Lovecraft. Nearer our own time the tragedy of Arthur Machen, a gifted fantasy writer who wrote exquisitely at his best, but who passed most of his professional life as a boozy and disillusioned reporter on the *London Evening News*. As my colleague Julian Symons has commented, he once in 1922 "totted up the moneys he had received in return for forty years' work in literature, and found that they came to a total of £635.00. This was his return for the production of eighteen books, which includes a translation of Casanova's Memoirs that ran to twelve volumes".

And in our own day that great friend and champion of fantasy and the macabre, August Derleth, who wore himself out in the service of literature. He once wrote to me—and he was astonishingly frank about the financial affairs of Arkham House—that he had spent all his life writing against sixty-day promissory notes. August was 62 and had a long history of heart-trouble which proved too much in the end.

Equally poignant was the more recent death at the age of 52 of Rod Serling who, almost single-handed had formed a great outlet for the adaptation of macabre and fantasy works into cinema with his famous television series *Night Gallery*.

Universal bought two of my own works and Rod Serling's adaptation of my 'Camera Obscura' was a faithful and brilliant translation of the ideas on paper into terms of the cinema. His was an inestimable loss to the genre in film, as August's was in literary terms.

One could extend the list almost indefinitely. The life of Tim White, the author of *The Sword in the Stone*, was almost a horror story toward the end. After Tolkien, his was perhaps the greatest talent in the traditional fantasy field.

These histories leave a poignant flavour. How different is the lot of the macabre/fantasy writer today. For with the paperback book as well as the hard-back; with the film as well as the television outlet, the opportunities have never been more favourable for the writer as well as for those who simply enjoy cold grue.

And with the renaissance of Arkham House under the outstanding direction of James Turner on the editorial side and Forrest Hartmann in its business direction, the world's greatest publishing house in the macabre and fantasy field will go on for all the foreseeable future.

This is comforting to know, for the writer's life is a lonely one facing the wall each day with the blank sheet of paper in the typewriter. Perhaps some of our own fears and fantasies are loosed on to the page, together with our fictional ones? The world of Darkness, Mist and Shadow and Darkness is a rich and strange one indeed and the practitioners of this curious art are not always themselves conscious of what they are creating. To any would-be writer in the field, I would give the sincere advice: be subtle. For with circumspection and discretion, the impact is far more powerful when the denouement comes.

I am nearly at the end, but to illustrate my point I cannot resist quoting the final paragraph of what I consider to be my best traditional ghost tale, 'The House by the Tarn'. It is simply this, "Down below a door slammed where there was no door and heavy footsteps, echoing where there was no floor, advanced towards him".

Consciously or not, there is a reflection here of the world's oldest macabre story, "A light was put into his hand". The fantasy writer is a dreamer and a poet as well as a businessman. Everyone in this room today is such a dreamer, united in the love of an art which we hold in such respect and with so much pleasure.

It is not for me to delineate my own role, but in my opinion the greatest British practitioners in the field of the traditional macabre story today are Roald Dahl, for the nasty sting in the tale; and Robert Aickman for the traditional ghost and macabre narrative. In the field of sheer fantasy Michael Moorcock is unparalleled on this side of the Atlantic just as Ray Bradbury and Robert Bloch reign supreme in the U.S.

Where is the genre going, you may ask? It has never been in better hands. There is now in this country a renaissance of excellent writers in the macabre/fantasy field. In the forefront are such prolific practitioners as Ramsey

Campbell, Kit Pedler, Brian Lumley and Richard Davis, to mention only a few. I am delighted to have made the acquaintance of some of them here at this Congress for the first time.

In the field of the anthology much has been done and will continue to be done by such editors as Richard Davis, Michel Parry, Peter Haining and Hugh Lamb, again to mention only those which spring readily to mind. It is difficult to overestimate the debt the writer owes to such editors, without whom it would be extremely difficult to find ready outlets for a single story.

Together with Arkham House on the other side of the Atlantic, they and their colleagues will ensure a stronger and brighter future both for the writer and for the enthusiast of the macabre.

I am closing now, but before I do so I would like to record my grateful thanks; not only to the President, Chairman and officials of the British Fantasy Society who have kindly invited me here; but to the critics, fellow writers, delegates and to the readers and enthusiasts for macabre and fantasy fiction for their kindness, courtesy and love for the genre. Without the readers we, the writers, would not exist and I think I express the gratitude of all the professionals here when I say what an honour, privilege and pleasure it is to meet you all face to face.

Let us all go forward through the land of Darkness, Mist and Shadow. The best is yet to be.

—Basil Copper
February 2-4, 1977

BASIL COPPER became a full-time writer in 1970. His first story in the horror field, 'The Spider', was published in 1964 in *The Fifth Pan Book of Horror Stories*, since when his short fiction has appeared in numerous collections and anthologies, and been extensively adapted for radio and television. Along with two non-fiction studies of the vampire and werewolf legends, his other books include the novels *The Great White Space*, *The Curse of the Fleers*, *Necropolis*, *The Black Death* and *The House of the Wolf*. Copper has also written more than fifty hardboiled thrillers about Los Angeles private detective Mike Faraday, and has continued the adventures of August Derleth's Sherlock Holmes-like consulting detective Solar Pons in several volumes of short stories and the novel *Solar Pons versus The Devil's Claw*.

STEPHEN JONES lives in London, England. He is the winner of three World Fantasy Awards, four Horror Writers Association Bram Stoker Awards and three International Horror Guild Awards as well as being a seventeen-time recipient of the British Fantasy Award and a Hugo Award nominee. A former television producer, director and genre movie publicist and consultant, he is one of Britain's most acclaimed anthologists of horror and dark fantasy with more than ninety-five books to his credit. You can visit his web site at: *www.herebedragons.co.uk/jones*